A Catalogue of Manuscripts in the Entomology Library of The Natural History Museum, London

The Natural History Museum

A Catalogue of Manuscripts in the Entomology Library of The Natural History Museum, London

Julie M.V. Harvey, Pamela Gilbert and Kathy Martin

MANSELL

Historical Studies in the Life and Earth Sciences No. 5

First Published 1996 by
Mansell Publishing Limited, *A Cassell imprint*
Wellington House, 125 Strand, London WC2R 0BB, England
215 Park Avenue South, New York, New York 10003, USA

British Library Cataloguing-in-Publication Data
A catalogue record for this book is available
from the British Library.

Library of Congress Cataloging-in-Publication Data
Natural History Museum (London, England). Entomology Library.
Catalogue of manuscripts in the Entomology Library of the
Natural History Museum, London / Julie M.V. Harvey, Pamela Gilbert,
and Kathy Martin.
p. cm. – (Historical studies in the life and earth sciences : 5)
Includes bibliographical references and indexes.
ISBN 0–7201–2290–2
1. Entomology – Manuscripts – Catalogs. 2. Entomology – History –
Sources – Bibliography – Catalogs. 3. Manuscripts – England – London –
Catalogs. 4. Natural History Museum (London, England). Entomology
Library – Catalogs. I. Harvey, Julie M.V. II. Gilbert, Pamela.
III. Martin, Kathy. IV. Title. V. Series: Historical studies in
the life and earth sciences: no. 5.
Z6611.Z65N38 1996
[QL463]
016.5957–dc20 95–20562
CIP

Printed and bound in Great Britain

Contents

Acknowledgements

The authors would like to acknowledge the help given by many past and present members of the Department of Library and Information Services, The Natural History Museum. In particular we thank the current Head of Library Services Mr R. E. R. Banks and previous Librarian Mr A. P. Harvey for their support in this project. Mr G. Moody and Miss I. Gaster as past members of staff contributed to the early ground work on the catalogue. Mrs C. Gökçe, Miss A. Lum, Mr N. Stead, Mr R. Steers and vacation students over the years have all aided the completion of this project. Members of the Department of Entomology have helped with the identification of particular items and with establishing the relationship of various manuscripts with the scientific collections. We also acknowledge the assistance of Dr Peter Barnard, who kindly edited the scientific names in the manuscript.

Finally we offer our special thanks to Mrs Gail Fordham who has typed most of the catalogue onto disk.

Introduction

History of the Manuscript Collection

In April 1881 the British Museum (Natural History), now The Natural History Museum, opened its doors to the public for the first time. The zoological and botanical collections, which had previously been housed at the British Museum in Bloomsbury, had outgrown the available space. It was therefore decided that a new museum should be built at the South Kensington site to rehouse the vast natural history collections.

The complex planning for the transfer of all the natural history scientific collections to the new building unfortunately did not include the extensive natural history libraries. Indeed, over a period of years the original Alfred Waterhouse plan had been so eroded as to leave no space for a central library. Only a few published works which were heavily used by the scientific staff were transferred in 1880 together with duplicate items and selected works which were closely associated with the specimen collections.

During the original transfer few manuscripts and original drawings came to The Natural History Museum, although some manuscripts were later moved, following specific requests from the Keepers of the various Departments. Other manuscripts accompanied the scientific collections during the move as they contained important information relating to the specimens. For example, the 17 manuscript catalogues of the Sir Hans Sloane (1660-1753) collection were transferred, the two insect volumes are now housed in the Entomology Library.

After 1881 the zoological collections quickly increased in number. Specimen collections were acquired by purchase, presentation and bequest.

Many of these collections were accompanied by manuscripts, including collecting notebooks, diaries, correspondence and photographic collections. These manuscripts were either kept with the 'parent' specimen collection or separated and deposited in the Library.

The Zoological Library originally had full responsibility for entomological literature. After 1913 a separate Department of Entomology was founded but the Department of Zoology continued to control the acquisition of entomological publications until 1936. In 1975 another organizational change brought the management of all the departmental libraries and other centralized library services into one department, now known as the Department of Library and Information Services.

Over the decades the entomological manuscript collections gradually became dispersed across the Department of Entomology. Many items were uncatalogued and unlabelled and 'had found their way' into various cupboards, desks and filing cabinets. Few people knew of the existence and contents of these unique items.

In the 1970s the previous Entomology Librarian, Pamela Gilbert, OBE, made a determined effort to bring all the manuscript collections together in the Library. By 1980 nearly all the collections had been located and it was decided to start work on a detailed catalogue of the manuscript collection. This work was continued by the current Librarian, Julie Harvey, assisted by Kathy Martin. The project has been painstaking, as many items required research to establish the authorship of the work and the relationship between a manuscript and a particular scientific collection.

Manuscripts are still acquired today and so the collection continues to grow. Items are presented to the Library and manuscripts of notable scientific importance are occasionally purchased.

Size and contents of the collection

The Entomology Library contains over 350 separate manuscript collections. The size of each collection varies considerably. For example, small collections such as the Henry Walter Bates (1825-1892) collection of observations and notes made during his travels on the Amazon, 1851-1859, consists of two small notebooks. Large collections, like the Henry Stainton (1822-1892) correspondence, contain over 12,000 letters from British and foreign entomologists.

The collection spans over four centuries of entomological work and represents an important and rich resource for those researching the history of entomology and aspects of social history. One of the earliest items is a notebook by Charles Dubois (1656-1740). Although Dubois is better known as a botanist his notes and observations on collecting insects in London between 1692 and 1695 represent one of the earliest entomological manuscripts in existence.

Other collections of early historical importance include the folio letter book of Dru Drury (1725-1803), containing copies of letters made by Drury to important entomologists between 1761 and 1783, and the Sir Hans Sloane catalogue, which lists the insects in the Sloane collection, including specimens from Joseph Dandridge (1664-1746) and James Petiver (c. 1660-1718).

The nineteenth and twentieth centuries are well represented in the collection including works by eminent entomologists, such as William Kirby (1759-1850) Thomas de Grey, Lord Walsingham (1869-1919), Alfred Russel Wallace (1823-1913), Nathaniel Charles Rothschild (1877-1923), Richard South (1846-1932) and Owain Richards (1901-1984).

The majority of the manuscripts are those of British entomologists. Important foreign collections include works by Jean B. A. D. Boisduval (1799-1879), Louis Chevrolat (1799-1884), Harrison Dyar (1866-1929) and Philip Charles Zeller (1808-1883). Also many entomologists from around the world are represented in the various correspondence collections.

The many diaries and notebooks made by entomologists in Britain and abroad make fascinating reading and provide an insight into the character of collectors and of the places they visited in times past. Such collections include the diaries of Mervyn Palmer (1879-1955) recording his collecting trips in South America, and notebooks by Lucy Evelyn Cheesman (1881-1969) of her journeys to New Guinea, the New Hebrides and other Pacific islands. Many manuscripts were written by men serving in the armed forces while on duty in foreign lands. Such examples include the writings of Frederick Bailey (1882-1967), collecting in Nepal and India, and Thomas Bainbrigge Fletcher (1878-1950), who served on various warships, collecting insects while in port around the world. For those interested in the geographical aspects of the manuscripts, the index (pp. 247-49) should prove particularly useful.

A collection of particular note, recently acquired by the Museum, is the Janson Company archive. For over 90 years the family were important natural history agents, booksellers and publishers, based at Great Russell Street, London. This remarkable archive gives a unique insight into the extensive trade in natural history specimens during the late nineteenth and early twentieth centuries. The archive includes correspondence between the Janson family and professional collectors, institutions and private individuals from around the world. Such agents were responsible not only for providing the financial framework in which professional collectors would operate but also played an important role in distributing insect specimens and large collections across the world.

The catalogue lists a few manuscripts and archives produced by the staff of the Museum, including items of particular scientific and historical importance. For example, George Samouelle (d. 1846) was employed in the Natural History Department of the British Museum in Bloomsbury. He kept careful notes about entomological collections received by the Museum during the period 1821-1840, forming three volumes known as *Entomological Memorandums*. Another similar collection is known as *Insect Room Lists*. This consists of notes made by a number of staff about collections which were acquired by the Natural History Museum between 1879 and 1914. Each item from this work has been separately listed in this catalogue under the author's or collector's name. Important manuscripts in this work include notes on the insects collected by Charles Darwin (1809-1882) during the voyage of HMS *Beagle*, in the hand of Syms Covingdon, Darwin's amanuensis.

It should be noted that scientific correspondence produced by the staff of the Department of Entomology is deposited in the Museum archive and does not form part of this catalogue. Enquiries regarding the archive collection should be addressed to the Museum Archivist.

All collections which have been accessioned as manuscripts in the Entomology Library before 1995 are listed in this catalogue. They include standard manuscript collections, such as diaries, notebooks and correspondence collections. Some collections also include examples of nature prints. These were made by carefully transferring the wing scales of butterflies and moths onto gummed paper. Where a manuscript collection includes drawings, these have also been listed and described. Collections of drawings which include only minor notes or annotations have been excluded and will be listed in a future catalogue of the drawings collections.

Scientific significance of the collections

The historical importance of the collections has already been discussed. However, of paramount importance is the scientific data contained in the manuscripts, which include collecting data, details of localities, notes on insect food plants, observations about the taxonomy and general biology of insects. Many of the manuscripts are an integral part of the scientific specimen collections held in the Department of Entomology.

The detailed notes and observations made by entomologists in particular localities present an important record for biologists and conservationists today. For example, the Stanley Morris natural history diaries record detailed observations on all aspects of natural history from West Sussex between 1889 to 1942.

Other collections, such as the T. R. D. Bell (1863-1948) manuscripts, contain important information on the life histories of Lepidoptera. Bell lived at Karwar, Bombay, India and over a period of years bred many Lepidoptera specimens from larvae he had collected in the field. The manuscript details important descriptions of the larval, pupal and adult stages of Lepidoptera specimens, many of which were sent to The Natural History Museum.

Explanation of catalogue entry

Sample entry

287. SAUNDERS, Edward (1848-1910)

Saunders studied the Buprestidae of the world and visited all the main European museums which had collections of Buprestid type material. In 1871 he published *Catalogus Buprestidarum synonymicus et systematicus.* Later he studied and published papers on British Hemiptera and the Aculeate Hymenoptera. In 1892 he published *The Hemiptera Heteroptera of the British Islands* and in 1896 *The Hymenoptera Aculeata of the British Isles.* He was a Fellow of the Entomological Society, the Linnean Society and the Royal Society.

Biographical Reference: 1910. Anon. *Entomologist's Record and Journal of Variation* **22**:75-76.

Collection

(1) One Ms notebook: *Notes on Buprestidae.* Contains notes on Buprestidae found in collections during his European tour.
(2) One Ms notebook: *Register of Buprestidae,* 1872.

Explanation

The following abbreviations are used throughout the catalogue:
Ms – manuscript
Mss – manuscripts
Ts – typescript
Tss – typescripts

Catalogue number: A catalogue number is assigned to each collection in this work. This number is referred to in the name and geographical indexes.

Main heading: the main heading for each collection is usually the author of the manuscript(s) or the name most closely associated with the work. For example, the collection of 14 letters sent to Robert Adkin (1849-1935) are listed under the *recipient*'s name. The name given in the main heading is listed in as complete a form as possible. All biographical dates are shown throughout the catalogue in brackets, as are dates of publication, but other dates are not bracketed.

Biographical notes: Brief biographical notes for each main author are given, when available. These include details of notable publications, particular entomological interests, expeditions, career history and membership of learned societies.

Where possible, The Natural History Museum accession number of specimen collections relating to a particular manuscript collection have been given. These are indicated by the initials B.M. (British Museum) followed by an accession number, e.g. B.M.1948-41.

Biographical reference: One or two selected biographical references are cited to provide additional sources of information for researchers.

Collection: The details of each manuscript collection are listed in a standard format. Each distinct set of items within a collection is given a number and these sets may be further subdivided using roman numerals. The arrangement of these items reflects the sequence of the holdings on the shelves in the Library.

The number is followed by a brief description of the format, e.g. one Ms notebook, or one loose-leaf folder. After a colon the title of the work is given. The full wording of titles appearing on the manuscript have been transcribed in full. Titles which have been assigned to unmarked collections are given in square brackets. Where necessary explanatory notes have been added.

All collections are dated from the manuscript itself or have been given approximate dates by the compilers.

All correspondents in small collections of letters are normally listed in full, and the dates of the letters are noted in the right-hand column. It has not proved possible to list all the correspondents in large collections. In such cases the compilers have included only entomologists who, in their opinion are of particular importance, or who were closely associated with the recipient. It is hoped that in the fullness of time detailed finding aids to individual collections will be made.

Indexes

Name index

This refers the reader to names which appear as main headings for a collection and to names which are noted within a collection, such as correspondents or collectors.

Geographical index

The geographical index provides readers with a means of finding manuscripts relating to particular countries or regions. It refers only to countries mentioned in the sections headed *Collection*.

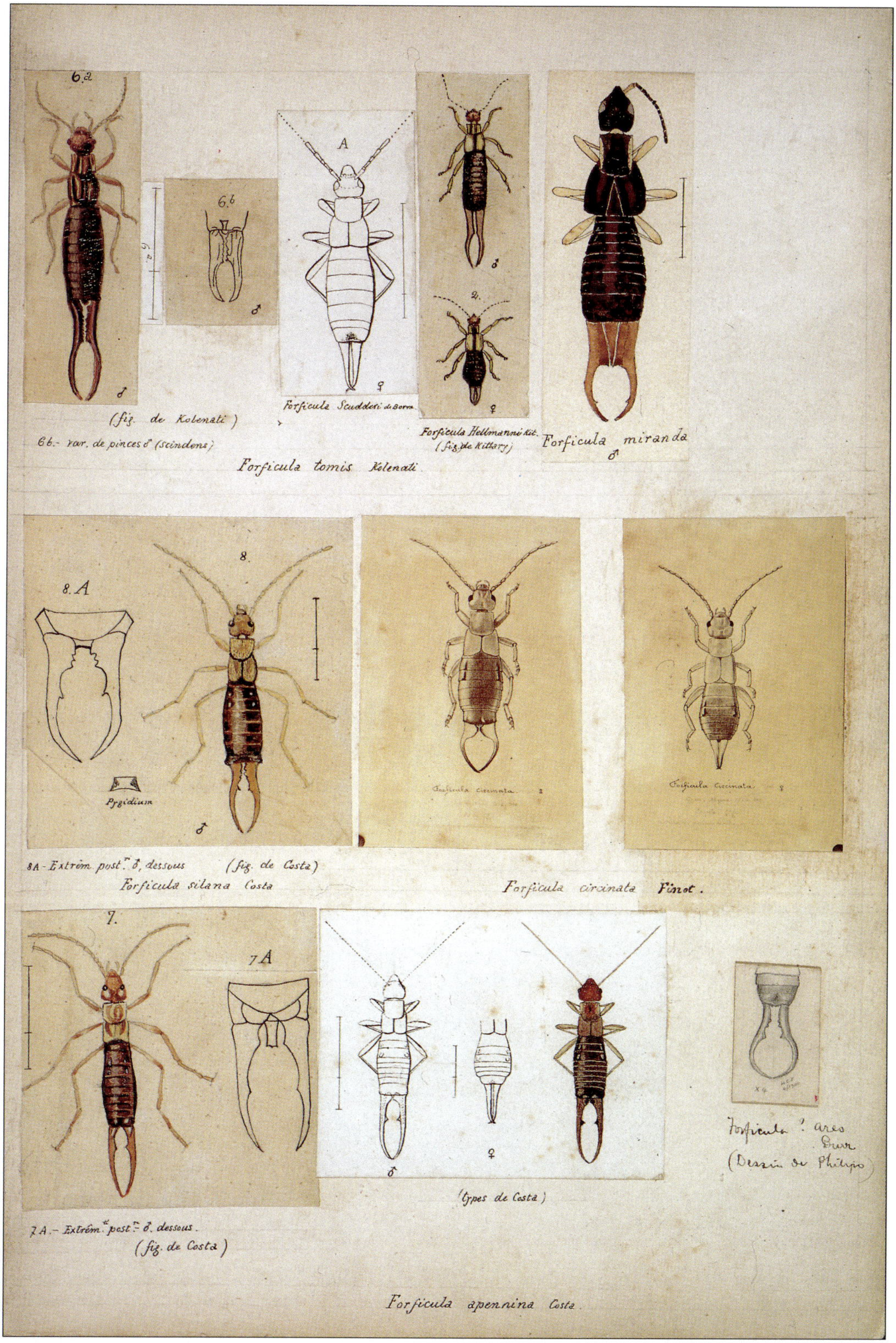

Auguste de Bormans was an officer in the French army. He became an acknowledged expert on the earwigs (Forficulidae). His manuscript collection includes many fine watercolours and detailed taxonomic drawings (see collection number 49).

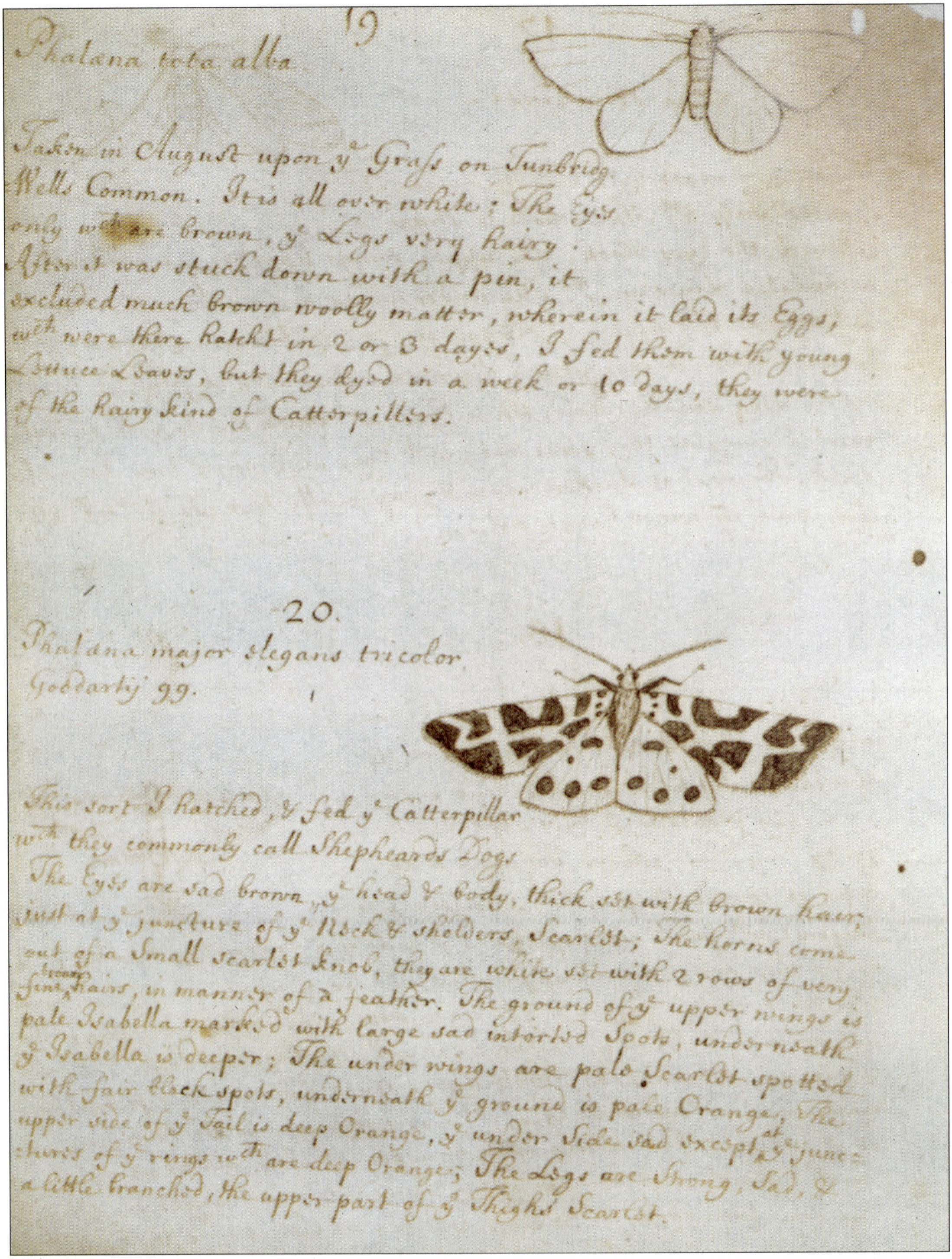

19

Phalana tota alba.

Taken in August upon ye Grass on Tunbridg Wells Common. It is all over white; The Eyes only wch are brown, ye Legs very hairy. After it was stuck down with a pin, it excluded much brown woolly matter, wherein it laid its Eggs, wch were there hatcht in 2 or 3 dayes, I fed them with young Lettuce Leaves, but they dyed in a week or 10 days, they were of the hairy kind of Catterpillers.

20.

Phalana major elegans tricolor Goedartij 99.

This sort I hatched, & fed ye Catterpillar wch they commonly call Shepheards Dogs
The Eyes are sad brown, ye head & body, thick set with brown hair, just at ye juncture of ye Neck & shoulders, Scarlet; The horns come out of a Small scarlet knob, they are white set with 2 rows of very fine brown hairs, in manner of a feather. The ground of ye upper wings is pale Isabella marked with large sad intorted Spots, underneath ye Isabella is deeper; The under wings are pale Scarlet spotted with fair black spots, underneath ye ground is pale Orange, The upper side of ye Tail is deep Orange, ye under Side sad except at ye junctures of ye rings wch are deep Orange; The Legs are Strong, Sad, & a little branched, the upper part of ye Thighs Scarlet.

The Charles Dubois (1656–1740) diary represents the earliest manuscript held in the Entomology Library of The Natural History Museum. For the years 1692 and 1695 Dubois kept notes and made drawings of insects he had collected from South London (see collection number 120).

Rev. Arthur Miles-Moss lived in Pará, Brazil from 1911 to 1945. He was known as the vicar of the largst 'parish' in the world as his ministry covered the whole Amazon region. He was an accomplished artist, his drawings of larvae of exotic lepidoptera drawn c. 1922–1935 are considered of great scientific interest. The drawings are annotated with manuscript notes on the collecting and rearing of many South American species (see collection number 230).

Between the years 1851 and 1859 Henry Walter Bates (1825–1892) collected thousands of insect specimens from the Amazon region. During this time he kept a meticulous diary of his notes and observations which include many fine watercolour and pencil drawings of butterflies and beetles (see collection number 33).

The Catalogue

ANONYMOUS

1. One Ms notebook: *Catalogue of insects caught in South Africa particularly in the Natal area*, 1827-1832. Notebook also includes notes on tribal customs.
2. One Ms volume: *Catalogue of Hesperiidae (Suppl. 1887-1890).*
3. One typed Ms: *List of Siamese Coleoptera*, c.1930. Arranged in families according to *A General Textbook of Entomology* by A.D. Imms.
4. One Ms notebook: *Synonymic List of British Macrolepidoptera*, c.1912. Also inserted in notebook: *List of European books of entomological reference.*
5. One Ms diary: Includes natural history notes made during a visit to Japan, and lists Japanese insect specimens, 1876-1878.
6. One Ms notebook: *Insects and Disease Cases in the Central Hall*, 1901-1907. Concerns an exhibition in the Central Hall of The Natural History Museum. It identifies the model makers and catalogues the items on display.
7. One Ms notebook: *Excerpta e Museo Ludovicae Ulricae: continentes textum integrum de iis speciebus quae Majo 1814 in Museo nostro deficere videbantur, nec non observationes omnes, quas b. [J.C.W.] Illiger exemplari quo usus fuerat, inseruit, c.1814.*
8. Two Ms leaves: *A notice of collectors of British Insects with an account of the present state of their collections*, n.d. Presented by the Entomological Society in 1898.

9. ABBOT, John (1751-1840)

As a young man of 22 John Abbot decided to abandon his career in law and go to America where he would earn his living by collecting and painting insects. These drawings were sold to interested individuals in England and Europe. After a short time in Virginia, he finally settled in Georgia. His collections were in great demand and at a later stage he also collected birds and plants. He published nothing under his own name but his paintings were used and acknowledged by J. E. Smith (1759-1828) in *The natural history of the rarer Lepidopterous insects of Georgia* (1797). The watercolours he so painstakingly produced were accompanied by manuscript notes. Many included detailed information on the life histories, habitats etc. which were previously unrecorded.

Biographical References: 1984. Wilkinson, R. S. *Entomologist's Record and Journal of Variation* **96**: 110-123, 165-176, 222-229, 273-285; Gilbert, P. *John Abbot. Watercolours from Colonial America.* pp. (in press)

Collection

(1) 17 volumes: 2,758 original watercolours of American Lepidoptera with extensive Ms notes.
(2) Two volumes: 167 original watercolour drawings with extensive Ms notes (part of the Rothschild bequest).
(3) Two volumes: 148 watercolour paintings of moths (part of the Walsingham bequest). There are no notes with these paintings.

10. ADAMS, Frederick Charlstrom (d.1920)

F.C. Adams was elected a Fellow of the Entomological Society of London in 1877. He was a considerable authority on British Diptera and carried out most of his collecting in the New Forest. He presented his collection of some 9,000 specimens to The Natural History Museum, London in 1919. B.M. 1919-303.

Biographical Reference: 1920. Austen, E.E. *The Entomologist's Monthly Magazine* **56**:256.

Collection

(1) One Ms catalogue: *Catalogue of British Diptera in the collection of Fredk C. Adams, 50 Ashley Gardens, Westminster SW. Chiefly collected by himself in the New Forest [Hampshire]. With notes, localities and dates of capture &c.*, c.1895-1915. The numbers in the catalogue correspond to specimen labels.

11. ADAMS, Herbert Jordan (1838-1912)

A keen Lepidopterist, he developed an extensive collection comprising some 140,000 palaearctic and exotic Lepidoptera housed in 68 cabinets. The entire Adams collection was bequeathed to The Natural History Museum, London. B.M. 1912-399. He was a Fellow of the Royal Entomological Society and Royal Horticultural Society.

Biographical Reference: 1912. Anon. *The Entomologist's Monthly Magazine* **48**:243.

Collection

(1) Nine Ms volumes: *Catalogue of the Lepidoptera*. Lists specimens in the Adams collection.

12. ADAMS, William Henry

(1) One letter to C. O. Waterhouse, 1894, referring to a pickled locust from the Gold Coast (Ghana).

Insect Room Lists: Volume 1, item 29.

13. ADKIN, Robert (1849-1935)

He was a partner in the family tobacco firm of Adkin Bros. and a member of the Board of the Imperial Tobacco Company. His enthusiasm for entomology dated from his early youth. He was a Fellow of the Entomological Society of London and twice its Vice-President, President of the South London Entomological and Natural History Society and of the Eastbourne Natural History Society. For many years he was a member of the prestigious and exclusive Entomological Club. His entomological interests lay with the British Lepidoptera. His collection of over 80,000 Lepidoptera was bequeathed to The Natural History Museum, London B.M.1935-352; his excellent library was bequeathed to the Royal Entomological Society of London.

Biographical Reference: 1935. Sheldon, W.G. *The Entomologist* **68**:145-147, portrait.

Collection

(1) One bound volume: 14 letters from various correspondents addressed to Robert Adkin, c.1896-1917.

Bower, Benjamin A.	1896
Carrington, John Thomas (1846-1908)	undated
Cheape, C. H. M.	1915
Church, F.	1911
Craig, Sellar Gerard	1914
Frohawk, Frederick William (1861-1946)	1896
Gibb, Lachlan (1852-1922)	1917
McArthur, Harry (1846-1910)	1897
Ructon, W.	1909
Salvage, T. W.	1896
Vine, Arthur Charles (1844-1917)	1910

14. AFGHAN COMMISSION

Collection

(1) One Ms leaf: *Localities of insects collected [Dilaram] Afghan Commission*, 1884-1885.

Insect Room Lists: Volume 1, item 13.

15. ALSTON, Arthur Hugh Garfit (1902-1958)

Foremost a botanical systematist, Alston started his career working in the Herbarium at the Royal Botanic Gardens, Kew. After a brief period at Kew he was appointed systematic botanist to the Department of Agriculture, Ceylon. In 1930 Alston was employed by the Botany Department of The Natural History Museum, London. He was an extremely keen collector, travelling abroad every year, particularly to the Balkans and Spain.

Biographical Reference: 1958. *Nature, London* **181**:1306-1307.

Collection

(1) One collecting notebook: *Greek Macedonia 1932*. Refers to specimens in the collection B.M.1932-269.

16. ANDAMAN ISLANDS

Collection

(1) Two Ms leaves: *List of Coleoptera from the Andaman Islands*. Lists the specimens presented by Raphael Meldola (1849-1915).

Insect Room Lists: Volume 1, item 18.

17. ANDREWES, Herbert Edward (1863-1950)

Andrewes spent most of his career working for the Indian Forest Service, studying entomology as a passing interest. After his early retirement from the service he quickly developed his entomological studies and became the recognized world authority on Oriental Carabidae. His major publications include two volumes of *The Fauna of British India, Carabidae, Carabinae* (1929) and *Harpalinae* (1935). His collection of 40,000 beetles was presented to The Natural History Museum, London and his extensive library to the Royal

Entomological Society of London.

Biographical Reference: 1951. Britten, E.B. *The Entomologist's Monthly Magazine* **87**:64.

Collection

(1) Thirty-six cloth files and two notebooks: these consist of notes on Carabid collections and identifications from private and institutional collections that Andrewes had seen throughout the world.
(2) Four-volume loose-leaf catalogue: [Catalogue of Oriental Carabidae] lists collecting localities of material seen by Andrewes.

18. ANNANDALE, Thomas Nelson (1876-1924)

After a research fellowship in anthropology at the University of Edinburgh, Thomas Annandale went to India in 1904. For the rest of his working life he was to be associated with the Indian Museum of Natural History and then with its successor, the Zoological Survey of India. His early career was mostly concerned with anthropology, but later his interests widened to various aspects of zoology. From the time he joined the Skeal Expedition in 1899 he cultivated his interests in entomology. A glance at a bibliography of his papers shows the broad interest he had in all aspects of zoology.

Biographical Reference: 1925. Sen, S.K. *Records of the Indian Museum* **27**:1-28, bibliography and portrait.

Collection

(1) One Ms notebook: *Annandale Coll[ection]: Ruwenzori Coll.* Includes notes about material collected on the Ruwenzori (Uganda) Expedition 1905-1906 and unpublished notes and descriptions of new species.

19. ATKINSON, Edwin Felix Thomas (1840-1890)

Born in Tipperary, Ireland, he joined the Indian Civil Service in 1862. Atkinson was to become Accountant General of Bengal and President of the

Board of Trustees of the Indian Museum. During his lifetime he held several senior posts in the Indian Civil Service. He published a gazetteer to the North Western Province of India and was also the author of works on Indian law. His particular entomological interest was Indian Rhynchota. He published several entomological papers in the *Journal of the Asiatic Society of Bengal* and started *Indian Museum Notes,* a journal dealing mainly with Indian economic entomology.

Biographical reference: 1890. Walsingham. *Proceedings of the Entomological Society of London* 1890 :lix.

Collection

(1) One loose leaf Ms: *Notes on Indian Rhynchota.*

20. ATKINSON, William Stephen (1820-1876)

A very industrious collector and particularly of Indian Lepidoptera. He lived for many years in Calcutta collecting and painting specimens. His considerable collection was purchased on his death by W.C. Hewitson (1806-1878). Frederic Moore (1830-1907) published many of the new species found by Atkinson.

Biographical Reference: 1876. Westwood, J.O. *Proceedings of the Entomological Society of London* 1876 : xliv.

Collection

(1) One bound volume: 43 original drawings of Indian Lepidoptera [by native artists]. Many are annotated with collecting notes and the identifications are initialled by F. Moore (1830-1907).

21. AUSTEN, Ernest Edward (1867-1938)

Austen joined the staff of The Natural History Museum, London in 1889 and was given charge of the Diptera collections. In 1895-1896 he travelled as Naturalist on the cable ship *Faraday* and collected in Brazil. In 1899 he went with Ronald Ross to study mosquitoes in Sierra Leone. He eventually specialized in blood-sucking flies. His major publications include *A Monograph of the Tsetse-Flies* (1903); *Illustrations of British Blood-sucking Flies* (1906) and *Bombyliidae of Palestine* (1937).

Biographical Reference: 1938. Blair, K.G. *The Entomologist's Monthly Magazine* **74**:42-43.

Collection

(1) One loose-leaf Ms: [Unpublished records of African blood-sucking Diptera].
(2) Eight proof plates and correspondence with printers regarding the publication of *Oriental Blood-sucking Flies*.
(3) Four Ms notebooks:
  (i) Entomological notes compiled during a visit to Palestine, 1918.
  (ii) *Notes on Exotic Bombyliidae*, c. 1897.
  (iii) Two notebooks of miscellaneous notes and identifications of Diptera, sketches and newspaper cuttings.

22. BACOT, Arthur William (1866-1922)

Bacot's original interest lay with the Lepidoptera and he published many papers in *The Entomologist's Record and Journal of Variation*. In 1911 he was appointed Entomologist to the Lister Institute and in 1914-1915 acted as entomologist to the Yellow Fever Commission in Sierra Leone. In 1916 Bacot was honorary entomological advisor to the War Office and in 1917 joined the British Trench Fever Committee. He died of typhus in April 1922. His yellow fever work resulted in a report, published in 1916: *Report of the Entomological Investigation undertaken for the Yellow Fever (West Africa) Commission for the year August 1914 to July 1915*. Bacot was a Fellow of the Royal Entomological Society of London.

Biographical Reference: 1922. Prout, L.B. *The Entomologist* **55**:142-144.

Collection

(1) Correspondence collection: 16 letters from Bacot to J.C. Warburg, 1896-1901, concerning insect specimens.

23. BADGLEY, W. F.

Collection

(1) Three bound notebooks: *Lepidoptera of Assam*, c. 1900. Contain drawings, nature prints and Ms notes.

24. BAILEY, Frederick Marshman (1882-1967)

Bailey was a Lt.-Colonel in the British army, and also an explorer, linguist, secret agent and diplomat. He was the first to find the Himalayan blue poppy *Meconopsis betonicifolia baileyi*. He was a great collector of butterflies in Nepal and India but his collections extended to all areas of natural history. His many explorations gave him the material for *Mission to Tashkent* (1946) and *No passport to Tibet* (1957), well known books of their time.

Biographical Reference: 1967. *The Times, London,* 19 April.

Collection

(1) Seven collecting diaries: locality data for butterflies of Nepal and India, c. 1918-1937. These include notes on Lycaenidae, Papilionidae, Hesperiidae, Satyridae, Nymphalidae, Pieridae and Danaidae.
(2) Correspondence collection: approximately 25 letters including letters from

Pocock, Reginal Innes (1863-1947)	1941-1946
Rothschild, Lionel Walter (1846-1937)	1922-1923

25. BAKER, Charles Fuller (1872-1927)

Although primarily an entomologist, Baker held posts during his career as a botanist and agronomist. Most of his spare time was spent collecting and amassing his enormous collection of Hemiptera. He edited *Invertebrata Pacifica* and was instrumental in starting the *Pomona College Journal of Entomology.* In 1912 he became Professor of Agronomy at the University of the Philippines. He left the Philippines only once. Every moment of his spare time was devoted to the curation of his collection. This amounted to some 400,000 specimens and was undoubtedly the largest private collection covering the extreme western Pacific. His main insect collection was bequeathed to the US National Museum and his plant collection to the University of the Philippines.

Biographical References: 1927. Essig, E.O. *Journal of Economic Entomology* **20**:48-754, portrait; 1927. Hernandez, B. *Lingnan Journal of Science* **5**:271-279, bibliography only.

Collection

(1) Collection of Tss.
- (i) *Australian Cercopoidea.*
- (ii) *Some Philippine and Malaysian Machaerotidae (Cercopoidea).*
- (iii) *Tribes of Oriental Selenocephaloid Jassoidea, 1927.*
- (iv) *Australian Cercopoidea.*
- (v) *Jassoidea of Australia.*

(2) One loose-leaf Ms: *Homoptera sent to Mr.[W.L.] Distant.*

26. BALFOUR-BROWNE, John

Son of William Alexander Francis Balfour–Browne (1874-1967), he is also an enthusiastic coleopterist and was Head of the Coleoptera Section in the Entomology Department of The Natural History Museum, London prior to his retirement.

Collection

(1) One Ts: *The generic names of British insects: Palpicornia.*

(2) Ms notes and correspondence: concerns the Coleoptera collection of Edward Caldwell Rye (1832-1885), 1951.

(3) Four notebooks: *Identifications, station lists 1-355.*

(4) One Ms notebook: *South Africa, Field book of localities, 1954.*

(5) One Ms notebook: *List of stations; stock received*, 1954, South Africa.

(6) One Ms notebook: *Madeira, Field notes*, 1931.

(7) One Ms notebook: *Identification of insects from Azores and Madeira*, 1932.

27. BALY, Joseph Sugar (1816-1890)

A doctor of medicine by profession, his entomological interests were almost entirely Coleoptera, although he did a little work on the exotic Hymenoptera. Baly was a fellow of the Linnean Society and the Entomological Society of London.

Biographical Reference: 1890. Sharp, D. *The Entomologist* **23**:197-200.

Collection

(1) One Ms: *Index to the Baly Collection*, 23 pp.

28. BANKES, Eustace Ralph (1861-1929)

One of the great British collectors, he amassed nearly 100,000 specimens, mostly of microlepidoptera, which eventually became the property of The Natural History Museum, London. He wrote a large number of papers and contributed to such works as the *Victoria County History of Hampshire and the Isle of Wight*. His family home was Corfe Castle, Dorset, in which county he did much of his collecting.

Biographical Reference: 1930. Adkin, R. *The Entomologist* **63**:47-48.

Collection

(1) One Ms notebook: [Catalogue of lepidoptera captures], 1876-1880.

(2) One Ms: *Diary 1877-1885.*

(3) Six Ms diaries: *Entomological diaries and stray notes 1884-1911.*

(4) One Ms notebook: *Exchange book for specimens 1884-1910.*

(5) Ms notebook: I. *Lepidoptera of the Isle of Purbeck 1884-1902.* (Reference is made to Volume II, but this was not part of the collection when it came to the Museum).

(6) One Ms notebook: *Stray Notes on various Lepidoptera examined in different collections 1892-1911.*

(7) One diary: *Voyage to the Amazons on board Mssrs Siemens cable ship SS* Faraday, *1895-1896.*

(8) One notebook: [Digests of published papers by various authors.]

(9) One Ms volume: [Index to British Lepidoptera in *The Entomologist* 1864-1890 compiled by Grace Bankes and completed in 1905.]

(10) One Ms volume: [Index to the papers by Bankes published in entomological magazines in 1896.]

(11) One Ms volume: [An index of specific names which appeared only as generic names in *The Entomologist*, entries to the end of Volume 41, 1908.]

(12) One Ms notebook: *Index to the names of Lepidoptera – British and continental – mentioned in my various rough notes*, 1910.

(13) One Ms notebook: *Random extracts from notes left by Mr W. Machin in reference to the Lepidoptera in his collection which was sold by auction in 1895.*

(14) Ms notes: *E. Doubleday collection of Lepidoptera in the Bethnal Green Museum, 1894.*

(15) Ms notes: [Supplements to the Victoria Natural History of Hampshire.]

(16) Ms notes: [Lepidoptera of Wiltshire], 1899.

(17) Ms notes: *Lepidoptera taken or observed at and within a few miles radius of Salisbury, 1899.*

(18) Ms notes: c. 1903 for a contemplated *Catalogue of Dorset Lepidoptera* and draft lists for the *Victoria County History of Dorset*. (This was never published.)

(19) Ms notes: *Notes on some Haworthian [A.H. Haworth] specimens of British Lepidoptera.*

29. BANKS, Nathan (1868-1953)

Curator of Insects at the Museum of Comparative Zoology, Harvard and later Assistant Professor of Zoology, Harvard University. He was Vice-President and President of the Entomological Society of Washington. His major publications include, *A Treatise on the Acarina* (1904) and *A Catalogue of the Neuropteroid Insects (except Odonata) of the United States* (1907).

Biographical Reference: 1936. Wade, J.S. *Proceedings of the Entomological Society of Washington* **38**:104-105.

Collection

(1) *List of Neuroptera from North America received in exchange.*

Insect Room Lists: Volume 2, item 106.

30. BARNES, Thomas Alexander (1881-1930)

An extremely skilled collector and hunter who spent much of his life in Africa. Barnes made two expeditions on behalf of J.J. Joicey of the Hill Museum, Surrey, collecting Lepidoptera. He also made other trips hunting for big game specimens; one of the elephants in The Natural History Museum, London was collected by Barnes.

Biographical Reference: 1930. Talbot, G. *Bulletin of the Hill Museum* **4**:145-153.

Collection

(1) Correspondence collection: approximately 80 letters from Barnes to J. J. Joicey.

- (i) Letters from the first expedition, 1916-1920, to Belgian Congo.
- (ii) Letters, 1920-1929, includes second expedition to East Tanganyika, Kivu and the Lindi and Congo rivers.
- (iii) Four Ms leaves: List of correspondence, 1916-1921.

31. BARNS, H. B.

(1) One Ms notebook: Lists insects collected by Mr H.B. Barns, F. L. Hill and G.D. Wedd from the Monte Bello Islands, in 1952 on the ship HMS *Camparia*. B.M. 1953-11.

32. BATES, Frederick (1829-1903)

Born in Leicester and a brewer by trade, Frederick Bates was a brother of the celebrated H. W. Bates. He had the same passionate enthusiasm for natural history as his brother, his first love being Coleoptera. For a time his interests turned to freshwater algae and orchids, but eventually his interest returned to the Coleoptera. His collection was eventually purchased by The Natural History Museum, London.

Biographical Reference: 1903. Donisthorpe, H. *Entomologist's Record and Journal of Variation* **15**:347-349, bibliography.

Collection

(1) One folder:

- (i) One Ms notebook: *A Census of Mr. Fred Bates Collection of Heteromera with notes, October 1880.*
- (ii) Two Ms leaves: *List of the Halticidae as determined by myself (G.R. Waterhouse) and subsequently M. Allard.*
- (iii) Two Ms leaves: *British species of Halticidae in Hamlet Clarke's collection as determined by M. Allard.*
- (iv) Two Ms leaves: *British species of Halticidae selected from my collection to send to France, 1859.*

(2) One Ms volume: [Index to F. Bates Collection with localities]. This register came to The Natural History Museum, London, with the H. Donisthorpe Collection. It carries the signature of F. Bates and has Donisthorpe's note 'This register belonged to the late F. Bates brother of H.W. Bates'.

(3) One Ms notebook: *Catalogue of the duplicate Heteromera of the collection Dejean purchased of the Marquis F. de Brême by the Marquis de Laferté–Sénéctère 1841. Now in the possession of F. Bates Leicester 1867.*
(4) One Ms notebook: [List of *Eriodius* and *Tentyria*.]

33. BATES, Henry Walter (1825-1892)

Bates published his first notes on the Coleoptera in the *Zoologist* at the age of 18. He became acquainted with Alfred Russel Wallace at Leicester and they continued to correspond after Wallace left. In April 1845 Bates and Wallace departed for Pará in Brazil to collect insects and other natural history specimens. Bates lived in a very frugal way, having to sell his specimens in Europe in order to survive. He collected many new species. His contribution to the taxonomy of Lepidoptera was enormous. He also developed important ideas on the theory of mimicry. In 1864 Bates became Assistant Secretary to the Royal Geographical Society, and President of the Entomological Society of London. His *magnum opus, The Naturalist on the River Amazons* (1863) has been reprinted many times.

Biographical References: 1892. Sharp, D. *The Entomologist* **25**:77-80, portrait; 1949. Alston, A.G.H. *Geographical Journal*, London **112**:1-3, portrait.

Collection

(1) Two Ms notebooks: [The Amazon Expedition]. Contain important observations and notes by Bates as he travelled on the Amazon, 1851-1859. The notebooks are finely illustrated with many watercolour drawings of Lepidoptera and Coleoptera.

34. BAYFORD, Edwin Goldthorp (1865-1958)

A Fellow of the Royal Entomological Society of London for 51 years, and a long-standing member of the Yorkshire Naturalists' Union. As a youth he collected Lepidoptera, but his great enthusiasm was for Coccinellidae (Coleoptera). His collection is in the Birmingham City Museum. Bayford lived in Yorkshire all his life and was a member of the Yorkshire Dialect Society.

Biographical Reference: 1960. Uvarov, B.P. *Proceedings of the Royal Entomological Society* (C) **24**:52.

Collection

(1) Correspondence collection: Contains 466 letters mostly to E.G. Bayford, 1853-1920. Includes letters from:

Porritt, G.T. (1848-1927)	1897-1912
Roebuck, William Denison (1851-1919)	1895-1909
Sheppard, T.	1899-1905
Thompson, M.L.	1897-1903

These letters concern local natural history, the Yorkshire Naturalists' Union and the Yorkshire Coleopterists' Survey. They also cover the compilation by E.G. Bayford and others of the insect section of the *Victoria County History of Yorkshire.* The collection also includes many personal family letters.
(2) One typed index to the correspondence collection.

35. BELL, Thomas Reid Davys (1863-1948)

Bell lived at Karwar, North Kanara District, Bombay, India. During his time in India he bred many Lepidoptera specimens from larvae collected in the field. Many of the specimens were presented to The Natural History Museum, London. B.M.1934-394. He published on a variety of topics including a volume (1937) on the Sphingidae in the *Fauna of British India,* in collaboration with F.B. Scott.

Biographical Reference: 1948. Kinnear, N.B. *Journal of the Bombay Natural History Society* **48**:167-171, portrait.

Collection

(1) 68 Ms volumes, of which 63 are an unbroken sequence of diaries from 1884-1948. These include personal notes and natural history observations. There are three volumes of indices covering the period 1884-1910.
(2) One Ms notebook: A *List of named Indian Coleoptera sent to T.R.D. Bell,* c. 1908.

(3) Ts: [Notes on Indian Pyralidae], c. 1925.
(4) Six boxes of Ts: [Larval descriptions of lepidoptera reared in India, with determinations], c. 1900.

36. BENSON, Robert Bernard (1904-1967)

A member of the staff of the Entomology Department, The Natural History Museum, London, for some 40 years. His main interests lay with the sawflies which he collected extensively in England, on the continent of Europe and in Canada. He published a large number of papers and the three parts on British sawflies in the *Handbooks for the Identification of British Insects*. In addition to entomology, he had a considerable interest in the general conservation of wildlife.

Biographical Reference: 1974. Quinlan, J. *Bulletin of the British Museum (Natural History) Entomology* **30**:220.

Collection

(1) Six diaries: contain both personal and entomological data, 1922-1964, mostly United Kingdom.
(2) Five diaries: notes on special collecting trips:

- (i) Canada, 1956 .
- (ii) Austria, 1957.
- (iii) Sutherland and Wester Ross, Scotland, 1958-1961.
- (iv) Switzerland, 1959, 1962.
- (v) Canary Islands, 1964-1965.

37. BESSEMER, Henry Douglas (1895-1968)

A gifted amateur entomologist, by profession a chartered accountant. His extensive collection of some 40,000 butterflies was bequeathed to The Natural History Museum, London.

Biographical Reference: 1968. Anon. *Proceedings of the Royal Entomological Society of London* (C) **33**:53.

Collection

(1) One bound volume: *Butterfly and Moth Collection: Record of more important specimens with details of data and other material information. The numbers against the various specimens link up with the numbers against the insects in the cabinet.*
(2) One loose-leaf Ts: Records of specimens considered to be of most value and also the prices paid for purchased specimens.

38. BETTEN, Cornelius

Betten is best remembered as a trichopterist. He published a number of papers on the Trichoptera and collaborated with M. E. Mosely.

Collection

(1) One letter from C. Betten to M. E. Mosely, 1946.

39. BETTON, C Stewart

Collection

(1) Two Ms leaves: *List of localities* and *Contents of boxes, 1900.*
(2) One Ms leaf: [List of localities], 1902.
(3) Two Ms leaves: [List of localities and seasons], 1896-1897.

Insect Room Lists: Volume 1, items 4, 9, 11.

40. BIGNELL, George Carter (1826-1910)

Bignell spent his early years in the Royal Marines; when on shore posting he used his spare time to pursue his hobby of natural history. In 1864 he left the service and settled in Stonehouse, Gloucestershire, as Registrar of Births and Deaths. He became interested in and collected the parasites and hyperparasites of aphids. His collection of parasites and hyperparasites was unique and is now in The Natural History Museum, London.

Biographical Reference: 1910. Keys, J. H. E. *The Entomologist's Monthly Magazine* **46**:94-95.

Collection

(1) One Ms notebook: Insect numbers, with data and determinations. It refers to a collection of Hymenoptera. B.M.1907-10.

41. BINGHAM, Charles Thomas (1848-1908)

Bingham's career was with the Indian Army as a Lt.-Colonel in the Bombay Staff Corps. His first interest was ornithology. Later he turned to the study of Aculeate Hymenoptera, on which he worked, on his return to England after retirement. He succeeded Dr Blanford in editing the Hymenoptera and Butterfly volumes of the *Fauna of British India.*

Biographical Reference: 1909. *Journal of the Bombay Natural History Society* **19**:214-215.

Collection

(1) One Ms volume: *Synopsis of Genera: Butterflies of India.*

42. BLACKBURN, Thomas (Rev.) (1844-1912)

After a short career in the civil service, Thomas Blackburn decided to take holy orders. His interest in entomology was of long standing having contributed short papers to several entomological journals. He was for some time one of the original editors of *The Entomologist's Monthly Magazine.* In 1876 he went to the Hawaiian Islands as chaplain to the Bishop of Honolulu. During his residence there he took a great interest in the entomological fauna. A result of this was his paper with D. Sharp on the Coleoptera of the Hawaiian Islands. Later he went to South Australia and there he engaged himself in the study of the fauna. He wrote many papers on the Australian Coleoptera.

Biographical Reference: 1912. Lea, A. *Transactions of the Royal Society of South Australia* **36**:v-xi, bibliography, portrait.

Collection

(1) One annotated Ts: [Index of Coleoptera type specimens in the Blackburn collection].
(2) Two letters 1879 and 1880: concern Hymenoptera specimens sent to Frederick Smith, B.M.1879-40.

Insect Room Lists: Volume 2, item 92.

43. BLAIR, Kenneth Gloyne (1882-1952)

Blair was an outstanding naturalist, having an intense interest in all aspects of nature. In 1910 he took the post of Assistant in The Natural History Museum, London, where his chief area of responsibility was a large part of the Coleoptera collection. His work on the Coleoptera resulted in numerous publications. In 1932 he was appointed Deputy Keeper of Entomology. Ill health compelled him to retire in 1943, but he kept his enthusiasm and worked on the fauna of the Isle of Wight.

Biographical Reference: 1953. Hawkins, C.N. *The Entomologist's Monthly Magazine* **89**:25-27, portrait.

Collection

(1) 400 original pen-and-ink drawings by Barbara Hopkins. Drawn for an unpublished work by K. G. Blair, *Atlas of Coleoptera Larvae.*
(2) Three letters sent to Blair, 1931, from Cedric Dover, Forest Research Institute and College in Dehra Dun, India.

44. BLESZINSKI, Stanislaw (1927-1967)

A specialist working on Lepidoptera, particularly Crambidae and Pyraloidea on which he published over 50 papers. He worked in many of the entomological institutes of the world and was highly regarded by his peers. He was killed at an early age in a car accident.

Biographical Reference: 1971. Anon. *Entomologisches Nachrichtenblatt, Wien* **18**:33-36, bibliography and portrait.

Collection

(1) One Ts: Notes on the sub-family Crambinae from *Microlepidoptera Palaearctica* Bd. 1, 1965, ed. H. G. Amsel *et al.*

45. BODENHEIMER, Frederick Simon (1897-1959)

On graduating from high school in Germany in 1914, Bodenheimer was accepted into medical school at the University of Munich. At the outbreak of World War I he volunteered to join the army and was sent to the Eastern Front. After the war he obtained a PhD in applied entomology and decided to emigrate to Palestine in 1922. Publications included *Materialien zur Geschichte der Entomologie bis Linné* (1928-1929), *Die Schädlingsfauna Palästinas* (1930) and numerous papers on scale insects. In 1931 he was appointed Professor of Zoology at the Hebrew University in Jerusalem. His autobiography, *A Biologist in Israel,* was published just before his death in 1959. His publication list exceeds 400 titles, including several books. Bodenheimer travelled extensively and spent time as visiting professor at various universities including Minneapolis and Ankara.

Biographical References: 1959. Bodenheimer, F.S. *A Biologist in Israel: A Book of Reminiscences*. 492pp. Jerusalem; 1984. Harpaz, I. *Annual Review of Entomology* **29**:1-23, portrait.

Collection

(1) One Ts: *Scale Insects and Mealy Bugs (Coccoidea) of the Middle East*, 1945, with pen-and-ink original drawings.

46. BOETTCHER

Collection

(1) One notebook: [Localities of Philippine Coleoptera]. Contains a series of locality labels giving details of where and when specimens were collected. These have all been transcribed and given further details at the back of the volume.

47. BOHEMAN, Carl Heinrich (1796-1868)

Born in Sweden, he became a keen coleopterist and was an authority on the Chrysomelidae and Rhynchophora. Boheman published over 400 papers, describing many important economic insects. Important publications include *Insecta Caffraria annis 1838-45 a J.A.Wahlberg* (1848) and *Monographia Cassididarum* (1850-1862). He was a Fellow of the Entomological Society of London.

Biographical References: 1869. Stål, C. *Stettiner Entomologische Zeitung* **30**:35-38, 1931. Essig, E.O. *History of Entomology* vii, 1029pp. (pp.558-559, bibliography, portrait). New York.

Collection

(1) One Ms notebook: *Catalogue of Cassididae*, c. 1855 (original Ms of the work published in 1856: *Nomenclature of Coleopterous Insects in the British Museum*, Part 9: *Cassididae*).

48. BOISDUVAL, Jean Baptiste Alphonse Déchauffour de (1799-1879)

One of the founding members of the French Entomological Society and perhaps the most celebrated lepidopterist of France. Although his bibliography is not long, each work has a significant place in the literature of the subject. Among his most important papers are the entomological parts of the reports of the species collected during the voyages of *L'Astrolabe*, 1826-1829. His collection was a considerable size and went, during his lifetime, to Charles Oberthür. The bulk of the Oberthür collection was purchased later by The Natural History Museum, London and contains many of the Boisduval types.

Biographical Reference: 1880. Oberthür, C. *Annales de la Société Entomologique de France* (5)**10**:129-138.

Collection

(1) One Ms notebook: *Troisième cahier* (first and second are unknown). The notebook relates to the collection of Dr Boisduval. A note signed by René Oberthür, 12/3/1927, records 'this was given to M. [L.M.A.] Depuiset to accompany the collection'.
(2) One bound volume: 18 plates of watercolour drawings by E. Blanchard. Some of these drawings were used for Boisduval's Mémoire sur les Lépidoptères de Madagascar, Bourbon et Maurice, 1833, *Nouvelles Annales du Muséum d'Histoire Naturelle* **2**:149-270.

49. BORMANS, Auguste de

Originally an officer in the French army, his first interest was always with insects and particularly with the Orthoptera. This interest eventually settled on the Forficulidae and it was in this particular field that he became the acknowledged expert being 'recognized as the only and final authority.' He amassed an unrivalled collection, which was given to Malcolm Burr in 1900. He published several major works during his lifetime. His *magnum opus* was the classic work, *Forficulidae und Hemimeridae* (1900). *Das Tierreich*, Lief. 11.

Biographical Reference: 1901. Burr, M. *The Entomologist's Record and Journal of Variation* **13**:85-88, portrait.

Collection

(1) Three Mss in a paper folder: *Manuscrits de M. Bormans Forficulidae, appartenant à M. Malcolm Burr 24 Décembre 1900.* Contains:

- (i) *Tables dichotome des espèces.*
- (ii) *Genera Dermapterorum.*
- (iii) *Tierreiches Forficulidae.*

(2) One Ms: *Catalogue alphabétique des Forficulaires décrits par S. H. Scudder.*
(3) One loose-leaf Ms: *Collections de Dermaptères.*
(4) One Ms: *Dermaptères notes.*
(5) Two Ms leaves: *Notes diverses.*
(6) One Ms notebook: *Catalogue des Forficulidae 1898, 1899 chassé par A. de Bormans.*
(7) One Ms: *Dermaptères.*
(8) One Ms: *Catalogue des Dermaptères de Bormans 1900.*
(9) Approximately 500 pen-and-ink and coloured drawings of Dermaptera.
(10) File of correspondence includes letters from:

Burr, Malcolm (1878-1954)	1899-1900
Krauss, Herman August (1848-1937)	1897-1899

50. BORNEAU

Collection:
(1) One Ms notebook: Collecting localities Nos 1-3301 [Sarawak] c. 1910. There is no indication that it refers to an existing collection.

51. BÖRNER, Carl J. B. (1880-1953)

Collection

(1) One Ms leaf: a letter offering for sale a collection of Collembola to The Natural History Museum, London, 1906.

Insect Room Lists: Volume 1, item 31.

52. BOWER, B. A.

Collection

(1) Three entomological diaries: Daily records of entomological captures or sightings, mainly from South East England, North Kent, and Essex, July 1886-October 1912. The records also contain cuttings from *The Times* on the annual weather statistics for that period.

53. BOWES, Anthony John Lee (1913-1942)

An avid collector of Lepidoptera, Bowes was a pilot in RAF Bomber Command in World War II and was killed on active service, aged 28. Much of his collecting was done in South East England for which he kept detailed entomological records. He published several papers in *The Entomologist.*

Biographical Reference: 1943. de Worms, C.G.H. *The Entomologist's Record and Journal of Variation* **55**:42.

Collection

(1) Seven Ms diaries: collection records for 1930-1931 and 1935-1939.
(2) One Ms notebook: [List of the Lepidoptera of North East Kent].
(3) Three Ms notebooks: *Notes on British Lepidoptera,* 1930-1942.

54. BOWLES, Edward Augustus (1865-1954)

A keen botanist and ornithologist, and also a Vice-President of the Royal Horticultural Society, he had wide interests in natural history. Much of his work was on the determination of the early stages of Lepidoptera. His findings were often sent to J.W.Tutt (1858-1911) for inclusion in *A Natural History of British Lepidoptera* (1899-1909). Bowles was an enthusiastic grower of food plants for lepidopterous larvae.

Biographical Reference: 1954. Edelsten, H.M. *The Entomologist* **87**:148-149.

Collection

(1) One entomological notebook: c. 1887-1898.

55. BOYD, William Christopher (d. 1906)

A lepidopterist of some note and a consistent contributor to *The Entomologist's Monthly Magazine* for nearly 35 years.

Biographical Reference: 1907. Bankes, E.R. *The Entomologist's Monthly Magazine* **43**:16.

Collection

(1) Two Ms notebooks: lists of specimens with identifying numbers for the years 1859-1889.

56. BRIGHT, Percy May (1863- 1941)

Percy Bright was born in India, where his father was working as a missionary. His family returned to England when he was eight years of age and settled in Bournemouth. He became heavily involved in charitable work and was also Chairman of the London Missionary Society for several years, travelling abroad extensively on their behalf. He became a devotee of British Lepidoptera, in particular the Lycaenidae, and amassed a considerable collection. A member of the South London Entomological and Natural History Society from 1890, Bright was elected a Fellow of the Entomological Society of London in 1894. In 1938 he privately published with H.A. Leeds, *A Monograph of the British Aberrations of the Chalk Hill Blue Butterfly* Lysandra coridon (*Poda, 1761*). On his death, his large collection of Lycaenid butterflies was bequeathed to the South London Entomological and Natural History Society. His foreign butterflies were sold. Before his death he gave his collection of Heterocera to Lord Rothschild. This was incorporated into the collection at Tring Musem, and is now in the collections of The Natural History Museum, London.

Biographical Reference: 1946. Russell G.G.C. *Proceedings of the South London Entomological and Natural History Society* **1946:**1-4, portrait.

Collection

(1) 18 pattern plates for *A Monograph of the British Aberrations of the Chalk Hill Blue Butterfly* Lysandra coridon (*Poda, 1761*).

57. BROOKS, Cecil Joslyn (1875-1953)

A chemist by profession, he spent much of his time in the East, particularly Sumatra, where he collected extensively both botanical and zoological specimens. After his early retirement he was attached to The Natural History Museum, London, revising the Amathusiidae and other groups of Lepidoptera. His collection was divided between The Natural History Museum, London, and the Castle Museum, Norwich.

Biographical Reference: 1955. Gabriel, A.G. *The Entomologist* **88**:23.

Collection

(1) One Ms notebook: a record of collection data, 1921.
(2) Three Ms notebooks: containing descriptions and remarks on various groups of insects.
(3) Unpublished manuscript: *A survey of the genera constituting the family Amathusiidae*, includes drawings and a map.
(4) Six Ms notebooks: containing notes and references to the Brooks collection of Lepidoptera (Rhopalocera), mostly Amathusiidae.

58. BROWN, Eric Septimus (1912-1972)

Brown's early interests were with the aquatic Hemiptera, but later his interests turned to insect migration. In 1946 he worked briefly on locust control in Tripolitania and in 1951 he joined the pool of entomologists with the Commonwealth Institute of Entomology. With this organization he worked on the pests of tropical crops for many years in Africa, the Middle and Far East and many of the Pacific Islands. He was a prolific publisher of his results as his bibliography shows. His collections were bequeathed to the Hope Entomological Collections, Oxford, and his papers passed to The Natural History Museum, London, from the Natural Resources Institute in 1988.

Biographical Reference: 1973. Betts, E. *The Entomologist's Monthly Magazine* **109**: 65-71, bibliography, portrait.

Collection

(1) Eight boxes of reports and manuscripts:

Box 1:
(i) One loose-leaf Ts: *Entomological Research on Premature Nutfall of Coconuts in the British Solomon Islands. Entomologist's Progress Report No.4.*
(ii) One loose-leaf Ts: [*Anoplolepis* and nutfall].
(iii) One Ts: *Aircraft applications of Insecticides in East Africa* by D. Yeo and R. Foster, 1956.
(iv) Correspondence and notes relating to the Rennell Islands.

Box 2:
(i) Seven packets of Mss [On the morphological condition of the male *Eurygaster integriceps*] a) [Analysis] b) [Iran 1958-59] c) [Iran 1959-60] d) [Iran 1960-61] e) [Turkey 1958-59] f) [Turkey 1959-60] g) [Record sheets].
(ii) Loose Ts and correspondence relating to the Gilbert and Ellice Islands, c. 1956-1957.

Box 3:
(i) Ms notes: [Notes on *Anoplolepis*], c. 1956.
(ii) Ms notes: [Nutfall and crop sample plots], c. 1954-1956.

Box 4:
(i) Ms notes: Melittomma *in native plots*, 1952.
(ii) Ms notes: [Seychelles economic insects], c. 1952-1953.
(iii) Ms notes: [Experiments on trees and the effects of *Melittomma*], 1952-1953.
(iv) Ms notes: [Records of adult *Melittomma*], 1952.
(v) Ms notes: *Insecticide experiments*, 1952.

Box 5:

(i) Tss: *Determination of density of* Eurygaster integriceps *in overwintering area*. 1958-1959; 1959-1960; 1960-1961.

(ii) Ts: *Notes on overwintering areas*. Records the overwintering of *Eurygaster integriceps* and other insects in Turkey and Iraq, 1958-1960.

Box 6:

(i) Ts: *Agronomy Report* by A.H. Green, Solomon Islands, 1956.

(ii) One Ms leaf: [*Amblypelta* – Taxonomy].

(iii) Ms notes: Amblypelta *Cocoa Experiment, Kikum*.

(iv) Ts notes: [Taxonomy of Ants].

(v) Ms notes: *Fauna of Palm Crowns*.

Box 7:

(i) Loose-leaf Ms: [Examination of body fat contents of *Eurygaster integriceps*].

(ii) Ms notes: [Parasites of *Eurygaster*], c. 1960.

(iii) Ms notes [Notes on *Aelia* at Hamadan], c. 1958-1959.

Box 8:

(i) One Ts: *Report on experiments on machine spraying of coconuts against* Pseudotheraptus wayi.

(ii) Miscellaneous photographs: [Solomon Islands].

(iii) Ts: [Pests of Zanzibar], c. 1953-1956.

(iv) Miscellaneous graphs.

59. BROWN, G. Edward

Collection

(1) One loose-leaf Ms: *On the occurrence in this country of a species of migratory locust new to Europe and unnatural migrations in general, 1869.*

60. BROWN, Thomas

Collection

(1) One Ms leaf: [Coleoptera for and from the British Museum, 1893].

Insect Room lists: Volume 2, item 84.

61. BROWNE, F. G.

Collection

(1) One notebook: [Pencil drawings of beetles].
(2) Loose-leaf Mss: relates to his publication *Pests and Diseases of Forest Plantations and Trees*, 1968.
(3) Loose-leaf Ms: *African collecting localities.*
(4) Loose-leaf Ms: *Descriptions of species (unnamed coll. F.G. Browne) in Malaysia, c. 1963.*
(5) Loose-leaf Ms: *Miscellaneous notes and investigations.*
(6) Notes and correspondence relating to C.I.E. identifications, c. 1966-1979.
(7) Appoximately 300 letters relating to *Pests and Diseases of Forest Plantations and Trees* (1968) and identifications of Scolytidae, c. 1964-1984.
(8) One Ts: *Ceylonese species of Scolytidae,* n.d.
(9) Loose-leaf Ms: [Notes on the classification of Scolytidae].
(10) Loose-leaf Ms and drawings: [Scolytid larvae].
(11) One Ts: *Ceylonese species of Platypodidae.*
(12) Miscellaneous notes and drawings.
(13) Loose-leaf Ms: [Notes on *Xyleborus*].
(14) Loose-leaf Ms: [Data on size variation (African *Xyleborus* and Platypodidae)].
(15) Loose-leaf Ms: Species groups of *Platypus.*
(16) Loose-leaf Ms: [Notes on *Doliopygus*].
(17) Loose-leaf Ms: [Notes on Platypodidae].
(18) Loose-leaf Ms: *Genera of the* Xyleborina.

62. BRUNETTI, Enrico Adelelmo (1862-1927)

Brunetti had a dual interest in music and entomology. He had a good musical training and was a talented composer for band, orchestra and piano. In 1904 whilst on a musical tour of the Dutch East Indies, China and Japan he made an extensive insect collection. He later settled in Calcutta for 17 years. In 1921 he returned to Europe, spending his summers in England and winters in Paris and Brussels. Brunetti published many papers. The Imperial Bureau of Entomology employed him to identify specimens and he worked for long periods on British Diptera. His entire collection, some 80,000 insects (mainly Diptera) was left to The Natural History Museum, London, B.M. 1927-184. The bequest also included an extensive library.

Biographical Reference: 1927. R.S.W. *Records of the Indian Museum* **29**:287-296, bibliography.

Collection

(1) One bound Ms: *Catalogue of Australasian Diptera,* Calcutta, 1909.
(2) Correspondence between Major E.E. Austen and Mrs Cotton (executor of Brunetti's will) regarding the collection, 1927.
(3) One bound Ms: *Catalogue of African Diptera,* c. 1924.

63. BRYANT, Gilbert Ernest (1878-1965)

Bryant travelled extensively to most parts of the world, amassing a considerable collection of Coleoptera and writing numerous papers on Coleoptera from various zoogeographical regions. His collection of some 50,000 Coleoptera was presented to The Natural History Museum, London. Bryant was also well known as the creator of the Bryant Index of Country Houses, and was an expert on English porcelain – Chelsea in particular.

Biographical Reference: 1965. *The Times, London,* 24 March 1965.

Collection

(1) One Ms notebook:[Borneo Coleoptera with dates of capture], 1913-1914.
(2) One Ms notebook: [Insect collecting in Sarawak], 1913-1919. Presented in 1964.

64. BUCHANAN-WHITE, Francis (1842-1894)

Born in Perth, Scotland, Buchanan-White studied medicine at the University of Edinburgh. After travelling the continent for a year in 1866 he settled in Perth for the rest of his life. His entomological interests concentrated on the Lepidoptera and the classification of the Hemiptera. He was also an accomplished botanist. Buchanan-White published numerous scientific papers, many appearing in the *Scottish Naturalist*, *Journal of Botany*, and *The Proceedings and Transactions of the Perthshire Society of Natural Science*. He was a Fellow of the Royal Entomological Society and the Linnean Society.

Biographical Reference: 1895. Trail, J.W.H. *Annals of Scottish Natural History* **1895**:73-91, bibliography, portrait.

Collection

(1) Two Ms notebooks: *The Buchanan-White Collection*. Includes a list of contents by W.R. Dolling.

65. BUCKELL

Collection:

(1) Three Ms notebooks: *British Butterflies*, c. 1880s. Volume 1 not present.
  (i) *Vol.2 Argynnidae* [Nymphalidae].
  (ii) *Vol.3 Satyrids* [Satyrinae].
  (iii) *Vol.4 Coenonymphidi Anosudi* [Satyrinae and Danainae].

66. BUCKLER, William (1814-1884)

Born at Newport, Isle of Wight, Buckler was an artist by profession, exhibiting regularly at the Royal Academy. In 1848 he settled at Lumley House, Emsworth in Hampshire and became interested in entomology. For the last 25 years of his life he devoted his energies to describing and drawing the larvae of British Lepidoptera. The descriptions were published regularly in *The Entomologist's Monthly Magazine*. After his death, *The Larvae of British Butterflies and Moths* (1886-1901) was published in 9 volumes, edited by H.T.Stainton (1822-1892) and then G.T. Porritt.

Biographical Reference: 1884. Anon. *The Entomologist* **17**:47-48.

Collection

(1) One folder containing four letters including:
- (i) One letter from William Buckler to James Terry (1824-1906) dated 23 December 1880, concerning the disposal of his drawings and specimens after his death.
- (ii) Anonymous undated note, concerning the disposal of Buckler's specimens and drawings.

(2) One Ms notebook: contains entomological and fly-fishing notes, and sketches of landscapes and other subjects.

(3) One bound volume: contains 62 watercolour drawings of British species of *Eupithecia* made for the Rev. H. Harper Crewe (d. 1883), 1860-1879. Includes a note by Thomas de Grey Walsingham giving the history of these and other drawings by William Buckler.

67. BUCKLEY, W. (d. 1940)

Biographical Reference: 1945. Cockayne, E. A. *Proceedings of the Royal Entomological Society of London* (C) **9**:48.

Collection

(1) One Ms: *Buckley's collection of midges on cattle*, Kuala Lumpur (Malaya), 1935-1937.

68. BURGESS, James Alexander (1916-1977)

Educated at Dulwich College, he graduated from Kings College, London in 1939. During World War II he served in the Intelligence Corps and the Commandos. From 1947 to 1955 James Burgess served in the Colonial Service in Uganda, and during this time formed a large collection of Ugandan Lepidoptera, including many new species. He later held a number of government posts and worked as Chief Secretary to the Condominium of the New Hebrides from 1974-1977, at the same time forming an extensive collection of New Hebridean Lepidoptera. He had an extensive correspondence with R.F. Healey and on Healey's death the Burgess and Healey collections were deposited in the Saruman Museum in Sussex, England (later the National Butterfly Museum, but now closed and the collections dispersed). This collection of correspondence was purchased by The Natural History Museum, London in 1983.

Biographical Reference: 1979. Anon. *Aurelian* **1**:16.

Collection

(1) Correspondence collection : approximately 100 letters from Robert F. Healey to Burgess, c. 1950-1977.
(2) Correspondence collection: approximately 70 letters from various entomologists and dealers, c. 1960-1977, including:

Braine, J.W.P.	1963
Goodden, R.C.	1968
Zappalorti, Michael	1967

(3) 50 drawings: *Malayan* Charaxes *and* Ornithoptera.
(4) Miscellaneous notes.

69. BURMAN, J.H.

Collection

(1) Ms notes and pencil drawings of aphids.

70. BURR, Malcolm (1878-1954)

A specialist in European Orthoptera and later in his life a world expert on the Dermaptera; Burr was a man of many talents. He was trained as a geologist, but was also a journalist, author and at the time of his death, Professor of English at Istanbul University. His collection of Dermaptera came to The Natural History Museum, London, and his Orthoptera to the Hope Entomological Collections at Oxford.

Biographical Reference: 1954. Uvarov, B.P. *Entomologist's Record and Journal of Variation* **66**:231-232.

Collection

(1) One folder: *Catalogus Dermapterorum*. Contains typed notes and letters.
(2) One folder: *Fossil Dermaptera*. Contains typed notes and drawings.
(3) Two folders: [Notes on Dermaptera].
(4) Two folders: contain original coloured drawings of Dermaptera, plus several undated letters.
(5) 20 folders of Ts: *A Monograph of the Dermaptera*, 1915. The title-page bears the legend 'printed for the Ray Society' but it is unpublished. It is a spirit duplicate of a typescript and also contains a large number of very fine original coloured drawings of Dermaptera.

71. BURROWS, Charles Richard Nelson (1851-1936)

For some time the incumbent of the parish of Mucking, Essex, Lewis was a man of prolific talents. He was an expert microscopist and an advocate of the use of genitalia as a diagnostic feature in insect taxonomy. His other interests included colour photography and chemistry. His original intention was to be a chemist, but he decided on the Church. Working exclusively with Lepidoptera he was particularly associated with F.N. Pierce, providing him with material and drawings for his *Genitalia of the Group Geometridae of the Lepidoptera of the British Isles* (1914), but refused to allow his name to appear on the work. On his death his specimens came to The Natural History Museum, London.

Biographical Reference: 1937. Turner, H.J. *The Entomologist's Record and Journal of Variation* **49**:22-24, portrait.

Collection

(1) One Ms notebook: [Index to the Burrows collection of Geometridae].
(2) Two Ms leaves: [Specimens sent to J. H. Durrant for incorporation in the B.M. Collections], B.M.1916-60.

Insect Room Lists: Volume 2, item 108.

72. BUTLER, Arthur Gardiner (1844-1925)

At the age of 19, Butler joined the Department of Zoology at the British Museum, Bloomsbury, and in 1879 became Assistant Keeper in charge of the Arthropoda continuing in this post until his retirement in 1901. Butler published over 500 entomological papers, his main area of interest being the Rhopalocera and in particular the Pieridae and the Lycaenidae. His publications included; *Catalogue of Diurnal Lepidoptera of the Family Satyridae in the British Museum* (1868); *Catalogue of Diurnal Lepidoptera described by Fabricius in the British Museum* (1869) *and Lepidoptera Exotica* (1869-1874). A. G. Butler was a Fellow of the Royal Entomological Society of London, the Linnean Society and the Zoological Society of London.

Biographical Reference: 1925. Riley, N.D. *The Entomologist* **58**:175-176.

Collection

(1) One Ms notebook: *Notes on Fabrician Diurnal Lepidoptera*, 1868.
(2) One Ms notebook: *List of papers written and species described*, 1864-1888.
(3) Loose leaf Ms:[Descriptions of genera and species].
(4) Letters from A.G. Butler to H. Strecker in USA, 1878-1879. Photocopies.
(5) Correspondence collection: approximately 30 letters sent to A.G. Butler and G.R. Waterhouse (1879-1888) from:

Bouavia, George	1879
Bowrey, James John	1879
Buchecker, Heinrich (1829-94)	1862
Chevrolat, Louis Alexander Auguste (1799-1884)	1876
Clough, G.T.	1879
Davis, Theodor	1888
Edmonds, Thomas	1883
Fowler, William Weeks (1849-1923)	1884
Geoffroy Sainte-Hilaire, Isidore (1805-1861)	1847
Hobson, William	1882
Hutton, Frederick Wollaston (1863-1905)	1879
Jones, Albert Hugh (1840-1924)	1882
Morrison, Herbert Knowles (1854-1885)	1883
Oberthür, Charles (1845-1924)	1880
Peake, P. George	1881
Perry, M.	1881
Pryer, Henry James Strovin (1850-1888)	1879
Roberts, Rowland	1878
Swainson, William, (1789-1855)	1840
Walton, John (1784-1862)	1856
White, F.H.	1888

(6) Loose-leaf Ms: *Progress in the Collection of Insects in the British Museum since the Purchase by the Government of the Collection of Sir Hans Sloane in 1753*, c. 1890.
(7) Pencil portrait, 10cm × 17cm, of Francis Walker (1809-1874), drawn by Butler.

73. BUXTON, Patrick Alfred (1892-1955)

Trained in medicine, Buxton served during the First World War in Mesopotamia and Persia. After the War he was appointed as entomologist to the Palestine government for two years. Later he organized a medical and scientific expedition to Samoa, the resulting publications being, *Researches in Polynesia and Melanesia* (1927-1928) and *Insects of Samoa* (1927-1935). In 1925 he became Director of the Department of Entomology of the London School of Hygiene and Tropical Medicine, later holding the chair of Medical Entomology at London University. For many years his main interest was in the tsetse fly and he was a world authority on insect–human relationships. Buxton was a Fellow of the Royal Entomological Society of London, serving as President in 1942 and 1953-1954. He was elected a Fellow of the Royal Society.

Biographical Reference: 1956. Wigglesworth, V.B. *Biographical Memoirs of Fellows of the Royal Society* **2**:69-84, bibliography, portrait.

Collection

(1) Five loose leaf files: [Diptera reared from fungi], c. 1950-1953.

74. CAMERON, Malcolm (1873-1954)

Cameron started his working life as a naval surgeon after qualifying in medicine at the London Hospital. He collected Coleoptera at the various stations he visited. His collection of some 12,000 specimens was presented to The Natural History Museum, London, in 1936, B.M. 1936-555. He was the author of the five volumes on the Staphylinidae:Coleoptera (1930-1939) in the series, *Fauna of British India*. His collection of over 55,000 Staphylinidae was presented to The Natural History Museum, London, together with holotypes, paratypes and much unidentified material.

Biographical Reference: 1954. E.B. Britton. *The Entomologist's Monthly Magazine* **90**:290.

Collection

(1) One Ts: *List of Coleoptera observed in the Maltese Islands by Cameron and A. Caruana Gatte, 1907.*

(2) Five Ms notebooks: *Cameron European Coleoptera*, 1898-1907.

(3) One Ms notebook: *Catalogue of Mr M. Cameron's collection from the West Indies, 1908*. Includes the note: 'The series in this book are designated by the letters W.I. and were captured during the cruise of the *Indefatigable* in the West Indies between January and October, 1908.' Includes notes on localities and

methods of collecting the numbered specimens.

(4) Miscellaneous notes.

(5) Correspondence collection: correspondence to M. Cameron, 1930-1954, from British and European entomologists including:

Basilewsky, Pierre (b. 1913)	1931-1954
Benick, Ludwig (1874-1951)	1933-1947
Bernhauer, Max (1866-1946)	1936-1939
Brookes, A.E.	1945-1953
Cerruti, Marcello (1908-1978)	1945-1949
Cooper, Joseph Omar (1893-1972)	1935
DeSaeger, H.	1947-1950
Donisthorpe, Horace St John Kelly (1870-1951)	1944-1949
Fagel, G.	1950-1951
Gardiner, J.C.M.	1934-1936
Hull, Wilfred John (1893-1965)	1947
Hutchinson, G. Evelyn	1933-1934
Keys, James H (1855-1941)	1934-1939
Koch, C.	1934-1938
Last, H.	1943-1951
Louwerens, C.J.	1934-1950
Marshall, Guy Anstruther Knox (1871-1959)	1940-1942
Paulian, Renaud (b. 1913)	1938-1946
Pendlebury, Henry Maurice (1893-1945)	1934-1939
Richards, Owain Westmacott (1901-1984)	1938
Riley, Norman Denbigh (1890-1979)	1939-1949
Roche, P.J.L.	1943
Sachtleben, Hans (1893-1968)	1934
Sanderson, Milton W.	1945-1953
Scheerpelts, Otto (1888-1975)	1934-1949
Schouteden, Henri (1881-1972)	1947-1948
Sewell, R.B.S.	1934-1939
Steel, O.W.	1943-1953
Van Straelen, V.	1935-1949
Villiers, André (1915-1983)	1947-1953

75. CAMERON, Peter (1847-1912)

Most of Cameron's working life was spent employed in the dye industry and later in calico printing. His interest lay in the Hymenoptera and in spite of ill-health and extreme poverty, brought about by loss of work, he was to become a most prolific author. Perhaps his best and most important works were *A Monograph of the British Phytophagous Hymenoptera*, 4 volumes (1882-1893), published by the Ray Society and the Hymenoptera volumes of the *Biologia Centrali Americana*; Volumes 1-2, (1883-1900) and (1888-1900) respectively.

Biographical Reference: 1913. G.M.W. *The Entomologist's Monthly Magazine* **49**:20-21.

Collection

(1) Eight Ms leaves: [Entomological lists].
(2) Two Ms leaves: [Purchased Hymenoptera], 1905.
(3) Two Ms leaves: [Purchased Insects], c.1905.
(4) One Ms leaf: [Hymenoptera list], B.M.1906-264.

Insect Room Lists: Volume 1, items 22, 23,23A,33A.

(5) Eight leaves: [Type Collection], B.M.1909-182.
(6) Five Ms leaves : *Cameron's Diapriinae,* B.M.1910-1905.

Insect Room Lists: Volume 2, items 77, 83.

76. CANDÈZE, Ernest Charles Auguste (1827-1898)

Candèze was a pupil of Lacordaire, the great French naturalist, who may well have influenced Candèze in his choice of career as an entomologist. Candèze studied medicine at Liège and Paris, and it was at Liège that he came under the influence of Lacordaire.

Candèze specialized in the study of the Elateridae, publishing a four-volume work, *Monographie des Élatérides* (1857-1863). He formed several collections of Elateridae, but in later years turned to collecting Belgian Diptera. Candèze was the author of several novels, a keen horticulturalist and very fond of music. He was also devoted to photography and developed a hand camera at a time when this was practically unknown.

Biographical Reference: 1898. Lameere, A. *Annales de la Société Entomologique de Belgique* **42**:504-519, bibliography, portrait.

Collection

(1) One Ms notebook:[species described by Candèze].

77. CARPENTER, Geoffrey Douglas Hale (1882-1953)

Carpenter qualified in medicine, but early in his career decided to combine his qualifications with his passionate interest in natural history by turning to tropical medicine. He started work on the bionomics of *Glossina palpalis* and the relation of this insect to the disease of sleeping sickness. He was also well known for his work on the butterflies of Uganda and his experiments on mimetic associations. Much of his work was with Edward Poulton (1865-1943), whom he succeeded as Hope Professor at Oxford.

Biographical Reference: 1953. Riley, N.D. *The Entomologist* **86**:155-156.

Collection

(1) Typed Ms: [Reports on Norfolk Island (Pacific Ocean) collections]. This includes material collected by J. McCornish in 1939. B.M.1940-154.

78. CHALMERS-HUNT, J. Michael (b. 1920)

The author of many scientific papers on the Lepidoptera, Michael Chalmers-Hunt has also written several important reference works, including *Natural History Auctions: 1700-1971, a register of sales in the British Isles* (1976), and *Local Lists of Lepidoptera* (1989). From 1973 to 1985 he was editor of *The Entomologist's Record and Journal of Variation*. He is a Fellow of the Royal Entomological Society of London, and has twice been President of the British Entomological and Natural History Society.

Biographical Reference: 1989. Chalmers-Hunt, J.M. *Local Lists of Lepidoptera*, portrait (on dust jacket).

Collection

(1) Approximately 1000 record cards: indexed data for the *Butterflies and Moths of Kent*, published in *Entomologist's Record and Journal of Variation* 1960-1981. Includes unpublished additions.
(2) Entomological correspondence collection: approximately 2000 letters sent to Chalmers-Hunt, 1937-1984.
(3) General correspondence: approximately 1000 letters, 1972-1985, concerning editorship of *The Entomologist's Record and Journal of Variation.*

79. CHAMBERS, Victor Horace (1911-1984)

Born in Luton, Chambers was educated at Dunstable Grammar School and at Imperial College of Science, London. He worked at the Murphy Chemical Company as a chemist and later as product planning manager. He collected Hymenoptera, mostly from Bedfordshire, and published papers in various journals, including The Hymenoptera Aculeata of Bedfordshire in *Transactions of the Society for British Entomology* **9**:197-222. Chambers was a Fellow of the Royal Entomological Society and served on the Council, 1967-1970. His collection was presented to The Natural History Museum, London, B.M. 1984-357.

Biographical Reference: 1985. Else, G. and Fergusson, N.D.M. *Antenna* **9**:18-19, portrait.

Collection

(1) One loose-leaf Ms and correspondence: notes on various subjects, including pollination, and letters from staff of The Natural History Museum, London, c. 1930-1950.
(2) Five Ms notebooks:[Notes on Hymenoptera] 1929-1950.

80. CHAMPION, Harry G. (1891-1979)

Son of the entomologist G.C. Champion (1851-1927), he was educated at the Royal Grammar School, Guildford and the Universities of London and Oxford. From 1914 to 1940 he worked for the Indian Forestry Service, and from 1940 to 1959 was Professor of Forestry at the University of Oxford. He published many papers in the *The Entomologist's Monthly Magazine*, and from 1932 to 1979 was on the editorial staff.

Biographical Reference: 1980. Thompson, G.H. *The Entomologist's Monthly Magazine* **115**:93-94, portrait.

Collection

(1) One collecting notebook, c. 1920.

81. CHAPMAN, Thomas Algernon (1842-1921)

Chapman started his career as resident physician and surgeon at the Glasgow Royal Infirmary. Thought by some to be one of England's finest entomologists of the time, he was primarily a Lepidopterist. All his work was carried out in a very thorough manner and usually with excellent illustrations. He was one of the first Lepidopterists to appreciate the value of genitalia in determining closely related species.

Biographical Reference: 1922. Sheldon, W.G. *The Entomologist* **55**:44-48.

Collection

(1) One box: Pencil sketches of Lepidoptera, mainly larvae.
(2) Seven boxes: Photographs of wings and genitalia of Lepidoptera. These appear to be unpublished. Some of the photographs are by Frederick Clark.

82. CHAWNER, Ethel Francis (1866-1953)

Her first interest was in sawfly larvae, which she collected in the New Forest. The major part of her collection came to The Natural History Museum,

London, in 1929.

Biographical Reference: 1954. Benson, R.B. *The Entomologist's Monthly Magazine* **90**:46.

Collection

(l) One Ms notebook: *Notes on sawfly larvae, including doubtful and unknown species*, 1892-1929.

83. CHEESMAN, Lucy Evelyn (1881-1969)

Well known as an explorer, broadcaster and authoress as well as an entomologist, Evelyn Cheesman made numerous expeditions to New Guinea, the New Hebrides and other islands in the Pacific. She wrote many scientific papers and several books on insects and her travels. *Things Worth While*, (1957) is mainly autobiographical. She was associated with The Natural History Museum, London for many years as a voluntary worker.

Biographical Reference: 1969. Smith, K.G.V. *The Entomologist's Monthly Magazine* **105** : 217-219, bibliography, portrait.

Collection

(1) One box: Four expedition notebooks and diaries, including
  (i) *Collecting sites of L.E. Cheesman: local maps, notes on topography, and publications.*
  (ii) *New Hebrides* [Pacific Ocean] *notebook.*
  (iii) *Hymenoptera of New Guinea – 25 genera of Braconidae.*
(2) One Box: Three expedition notebooks and diaries:
  (i) *New Caledonia* [Pacific Ocean] *Entomological Register* 1949.
  (ii) *Ants of New Caledonia and Lifu Islands.*
  (iii) *Cyclops Mountains, 1936.*
(3) One box: Loose-leaf Ms notes, sketches and photographs from her expeditions.
(4) One box: Loose-leaf Ms notes, and sketches of Hymenoptera.
(5) One box: Photographs and maps of locations visited on her expeditions.

(6) One box: Correspondence to Miss Cheesman, including: Admiralty Naval Intelligence Division regarding use of her Ms gazetteer for intelligence work, 1953, and letters from:

Alexander, Charles Paul (1899-1981)	1945-1949
Balfour-Browne, William Alexander Francis (1874-1967)	1938-1961
Barnard, D.	1944
Belkin, John N. (1914-1980)	1958
China, William Edward (1895-1979)	1955-1960
Crossland, C.	1939
Dartington, P.J.	1950-1951
Diakonoff, Alexander Mikhailovich (1907-1989)	1948
Doncaster, John Priestman (1907-1981)	1961-1966
Evans, John William (1906-1990)	1950-1951
Froggatt, J.L. (1891-1975)	1947
Gardiner, J.S.	1931
Goodenough	1943-1948
Haddock, A.C.	1931-1935
Hill, C.M.	1932-1933
Hincks, Walter Douglas (1906-1961)	1935-1948
Hocking, Brian (1914-1974)	1956-1957
Huxley, E.	1943-1949
Krombein, Karl von Vorse	1949-1952
Lieftinck, M.	1936-1957
Riley, Norman Denbigh (1890-1979)	1933-1954
Steenis, C.G.G.H.	1952
Tate Regan, C.	1931

(7) One box: Correspondence to and from Miss Cheesman, and miscellaneous Ms notes. Correspondence includes letters from:

China, William Edward (1895-1979)	1955-1958
Doncaster, John Priestman (1907-1981)	1966
Perkins, John Frederick (1910-1983)	1968
Riley, Norman Denbigh (1890-1979)	1940

(8) One Ms notebook: [Pencil sketches of plants].
(9) One Ms notebook: *Notes on early voyages in the Pacific Ocean.*
(10) Two Ms volumes [Gazetteer of New Guinea].

84. CHEVROLAT, Louis Alexandre Auguste (1799-1884)

A founder member of the Entomological Society of France. For more than fifty years Chevrolat was involved in the study of entomology, making a particular study of the Curculionidae (Coleoptera). He published several monographs on the French fauna, describing many new species of Coleoptera and the fauna of other countries including the entomological fauna of Cuba.

Biographical Reference: 1884. Lefèvre, E. *Bulletin de la Société Entomologique de France* (6) **4**:clvi-clvii.

Collection

(1) One Ms notebook: *Catalogue des Longicornes (Cerambycites)* [sic] *de la Collection Chevrolat formeé de celle de Palisot* [de] *Beauvois* [A.M.F.J.], [Audinet-]*Serville* [J.G.], *Dejean* [P.F.M.A.], *(Latreille* [P.A.], *Humboldt* [F.H.A.von], *& Bonpland* [A.J.A.], *Cailliaud* [F.], *Olivier* [G.A.], *(Riche* [C.A.G.] *pour les Ind. or. et Madagascar, Durand pr. Tanger, Hoffmansegg* [J.C.] *-Portugal, Illiger* [J.C.W.], *Fabricius* [J.C.], *Desfontaines* [R.L.], *& Bosc* [d'Antic, L.A.G.]) *et de ce qu'il a pu acheter, échanger avec ses correspondants et les Museés de Londres, Berlin et Leyde &c, &c pendant 30 années.*
Copied verbatim from the original manuscript by E.W. Janson Sept. 1863.

85. CHITTENDEN, Frank Hurlbut (1858-1929)

Born in Cleveland, Ohio, Chittenden studied at Cornell University. In 1891 he was appointed assistant entomologist to the US Department of Agriculture. He was editor of *Entomologica Americana* and was a Fellow of the American Association for the Advancement of Science. He published many taxonomic and economic entomology papers especially on insects infesting stored products.

Biographical Reference: 1936. Wade, J.S. *Proceedings of the Entomological Society of Washington* **38**:109-110.

Collection

(1) Two Ms leaves: *List of Sphenophori presented to the British Museum by the United States National Museum* 1910. B.M.1910-153.

Insect Room Lists: Volume 2, item 90.

86. CHRISTY, William Miller (1863-1939)

An enthusiastic collector of insects, he designed his own trap, and was successful in securing a large number of rare and scarce specimens. He once owned part of Wicken Fen, which he made over to the National Trust. His collection is in The Natural History Museum, London along with part of his extensive library.

Biographical Reference: 1940. Edelsten, H.M. *The Entomologist* **73**:24.

Collection

(1) Four entomological diaries: collecting records for 1889-1902.

87. CLARKE, Cyril Astley (b. 1907)

Clarke studied at the University of Cambridge. A doctor by profession, but with a keen interest in Lepidoptera, his studies on genetics and mimicry in Lepidoptera contributed to a medical breakthrough concerning the Rhesus negative condition in the unborn child. His large collection demonstrating genetic studies is now in The Natural History Museum, London.
He was one time President of the Royal College of Physicians and of the Lepidopterist's Society. He was elected Fellow of the Royal Society.

Biographical Reference: 1977. Kendall, R.O. *The Lepidopterist's Society Commemorative Volume 1945-1973*, pp. 374 (pp. 81-82) portrait.

Collection

(1) One folder:
  (i) One Ts: *Operation Arion*, 1952.
  (ii) Notes from Captain Purefoy on breeding *[Maculinea] arion* sent to C.A. Clarke, November 1951.

(2) One Ts: *Guide to the Clarke/Sheppard/Turner Genetic Collection of Butterflies*. Originally produced in 1982, with further updates in 1987 and 1989.

88. CLUTTERBUCK, Charles Granville (1871-1957)

Educated at College School, Gloucester, he was later a solicitor in the same town. Clutterbuck's particular interest was in the Lepidoptera, especially microlepidoptera. He added over 40 new species to the Gloucester County list, and was a member of the Cotswold Naturalists' Field Club. He was a Fellow of the Royal Entomological Society.

Biographical Reference: 1958. H.C.H. *The Entomologist* **91**:52.

Collection

(1) One Ms notebook: *The Macrolepidoptera of Gloucestershire*, 1942.

89. COCKAYNE, Edward Alfred (1880-1956)

A doctor of medicine, his entomological interests centred on genetics and teratology on which he published a great many papers. He amassed a large collection of Lepidoptera aberrations which in later years he amalgamated in the Rothschild–Cockayne–Kettlewell collection, now in The Natural History Museum, London.

Biographical Reference: 1957. Riley, N. D. and H.B.W. *The Entomologist* **90**:78-80.

Collection

(1) Two photograph albums: mostly photographs of Lepidoptera, 1903-1907.

90. COLLENETTE, Cyril Leslie (1888-1959)

Collenette worked in The Natural History Museum, London as an Honorary Associate, and in his later years became an expert on the Lymantriidae on which he published many papers. In addition to the Lepidoptera he maintained a long interest in general natural history, particularly ornithology. During his life he travelled extensively. He participated in the St. George Expedition, 1924-1925, to the Matto Grosso.

Biographical Reference: 1959. Longfield C.E. *London Naturalist* **39**:136-138, portrait.

Collection

(1) Five notebooks: Data on the Lepidoptera collected on the Matto Grosso Expedition.
(2) One Ms notebook: *Descriptions of lepidopterous larvae and pupae, St. George Expedition.*
(3) Ts and correspondence: Malaise trap collecting and the P.E. Viette material.
(4) Correspondence collection: Includes letters to and from:

Bryck, Felix (1882-1957)	1953-1954
Diakonoff, Alexander Mikhailovich (1907-1989)	1953-1956
Pendlebury, Henry Maurice (1893-1945)	1931-1935
Pinhey, Elliot Charles Gordan (b. 1910)	1959
Roepke, Walter Karl Johann (1882-1961)	1946-1952
Schouteden, Henri (1881-1972)	1935-1936
Toxopeus, Lambertus Johannes (1894-1951)	1948

91. COLLIER, Alan Egerton (d. 1972)

Collier was especially interested in the varieties and genetics of British Lepidoptera. He was a successful breeder of aberrations. His collection was donated to The Natural History Museum, London in 1972. B.M.1972-588.

Biographical Reference: 1974. Lees, A.D. *Proceedings of the Royal Entomological Society* (C) **38**:58.

Collection

(1) Six collecting notebooks and diaries. All refer to the collection presented in 1972.

92. COLYER, Charles Norman (1908-1970)

Charles Colyer was educated at Queen Elizabeth's School, Barnet. He developed a particular enthusiasm for the Diptera, and especially the Phoridae, on which he was the recognized authority of his time. He published with Cyril O. Hammond, *Flies of the British Isles* (1951), which ran to several editions. He was a Fellow of the Royal Entomological Society. On his death his Diptera collections came to The Natural History Museum, London. B.M.1970-489.

Biographical Reference: 1972. Smith, K.G.V. *The Entomologist's Monthly Magazine* **108**:1-2, bibliography, portrait.

Collection

(1) Three boxes of photomicrographs: the collection is retained in Colyer's original order and labelled with details of equipment used etc.
(2) Four loose-leaf files: [photographs and manuscript text on Diptera].
(3) One box of entomological photographs: includes both glass slides and photomicrographs, retained in Colyer's order.
(4) Correspondence collection: 1948-1960, includes approximately 500 letters between Colyer and Hermann Schmitz (1878-1960).
(5) Collection of Mss on Phoridae by C.N. Colyer. Includes:
   (i) One notebook: *Keys to British species of* Megaselia *and* Plastophora.

(ii) Two notebooks: *Translation into English of Beyer's Keys to World Genera of Phoridae.*

(iii) One loose-leaf Ms: *Keys to British Phoridae. Except* Megaselia.

(iv) One loose-leaf Ms: *Keys to British Phoridae* – Megaselia.

(v) Three loose-leaf Mss: [Keys to Palaearctic Phoridae].

(6) Correspondence collection: Approximately 100 letters between C.N. Colyer and staff of The Natural History Museum, London, 1951-1970.

(7) Miscellaneous correspondence to and from C.N. Colyer c. 1950-1970. Includes letters from:

Hobby, Bertram Maurice (1905-1983)	1954-1957
Oldroyd, Harold (1913-1978)	1962

93. CORBET, Alexander Steven (1896-1948)

A biochemist by training, he was employed by The Natural History Museum, London, to work on the butterfly collections. His most important contribution to entomology was his joint work with H.M. Pendlebury, *The Butterflies of the Malay Peninsula* (1934). A manuscript for the second edition was completed by him shortly before his death.

Biographical Reference: 1948. Riley, N.D. *Nature, London* **161**:1003.

Collection

(1) One Ms volume: *Indo-Australian Lycaenidae.* Index to species.

(2) Two collecting notebooks.

(3) One Ms volume: *Index to names of Malayan Rhopalocera, trivial names.*

(4) One Ms notebook: *Fabrician names of Indo-Australian Rhopalocera,* 1941.

(5) One Ms notebook: *Keys for Malayasian Rhopalocera. For the second edition of The Butterflies of the Malay Peninsula.*

(6) One Ts: *A Synonymic list of the Butterflies of British Malaya, with particulars of their distribution and references to figures in W.L. Distant's Rhopalocera Malayana, (1882-1886),* 1942.

(7) One Ms notebook: *Distant's Rhopalocera Malayana, figures,* 1942.

(8) One Ms notebook: *Distribution of Malayan Rhopalocera.*

(9) One Ts: *A catalogue of Bombyces and Sphinges recorded from the Malay Peninsula.*

(10) Pen-and-ink drawings, photographic negatives and correspondence relating to the third edition of *Butterflies of the Malay Peninsula* (1978), revised by J.N. Eliot.

(11) 19 Ms notebooks: [Notes on Lepidoptera].

(12) Pen-and-ink drawings, b/w photographs of Lepidoptera.

94. COTTAM, R.

Cottam became a Fellow of the Royal Entomological Society of London in 1923.

Collection

(1) Three Ms volumes: *Faunistic lists of Rhopalocera of the Dutch East and West Indies*, 1940.

95. COTTERELL, G.S.(1896-1977)

Cotterell spent 45 years as an agricultural entomologist. He went to the Gold Coast (Ghana) in 1920 and also travelled to other countries, including Nigeria and Afghanistan, as an advisor on agriculture.

Biographical Reference: 1979. Harris, K.M. *Antenna* **3**:56.

Collection

(1) One loose-leaf folder: *Handbook of Insect Pests of the Gold Coast with Notes on Treatment*. Unpublished manuscript with notes on pesticide treatments and original drawings. Presented in 1978 by K.M. Harris.

96. COWAN, Charles Francis (d. 1988)

A Lieutenant-Colonel in the British Army, his main interest was the Lepidoptera of India and Malaya. *The Butterflies of the Malay Peninsula* (1934) by Alexander Steven Corbet (1896-1948) and Henry Maurice Pendlebury

(1893-1945) was based on Cowan's collection of butterflies. He was particularly interested in the Lycaenidae, and published various papers on this group. Cowan was a bibliographer of some skill and knowledge, and many turned to him for help over dating and issue of publications. In his later years he published many papers in the *Journal of the Society for the Bibliography of Natural History*.

Biographical Reference: 1989. Gardiner, B. *Bulletin of the Amateur Entomological Society* **48**:70-71.

Collection

(1) One loose-leaf Ms : *Odonata Records 1959-1974*. Lists sightings of Odonata in Great Britain.

97. COWLEY, John (1909-1967)

An amateur entomologist of independent means, his particular interest was in the Odonata. In later years he turned to Diptera. Cowley's extensive collection of Odonata and related indexes were presented to The Natural History Museum, London, B.M.1968-70.

Biographical Reference: 1968. Kennedy, J.S. *Proceedings of the Royal Entomological Society* (C) **32**:59.

Collection

(1) Ten boxes of drawings: approximately 1000 pen-and-ink drawings of Odonata.
(2) Correspondence and memoranda regarding the Cowley bequest.
(3) Correspondence collection: approximately 200 letters to and from Cowley. c. 1934-1958. Includes letters from:

Fraser, Frederic Charles (1880-1963)	1934-1955
Kimmins, Douglas Eric (1905-1985)	1934-1935
Longfield, Cynthia (1896-1991)	1934-1958

98. COX, Charles James (d. 1889)

A member of the Botanical Society of London and a Fellow of the Entomological Society from 1853 until his death in 1889, Cox's main interests were with the Lepidoptera and Scolytid beetles. He produced a number of papers and a textbook, *Our Common Insects: First Steps to entomology* (1864).

Collection

(1) Loose-leaf Ms: Includes a number of manuscript chapters on entomology for a projected but unpublished series *Elementary Lectures upon Natural History*. It contains a number of pen-and-ink drawings.

99. CRAWTHAY, R.

Collection

(1) Five Ms leaves: [List of localities for specimens collected in Basutoland and Transvaal], 1901-1902.

100. CROWLEY, Philip (1837-1900)

Born at Alton in Hampshire, he came from a Quaker family and was a partner in a brewing business. Crowley had a wide interest in natural history and developed fine collections of exotic butterflies and birds' eggs. He was a Fellow of the Linnean, Zoological and Entomological Societies and was a keen supporter of the Croydon Microscopical Club.

Biographical Reference: 1905. Anon. *The Entomologist's Monthly Magazine* **37**:49.

Collection

(1) One Ms volume: [Catalogue of Lepidoptera].

101. CROWSON, Roy Albert

Educated at Judd School, Tunbridge and Imperial College, London University, he was appointed Curator of Tunbridge Wells Museum in 1938. In 1948 he became Lecturer in Zoological Taxonomy at Glasgow University, becoming Senior Lecturer in 1964 and Honorary Lecturer in 1980. Crowson published over 100 monographs and papers on various aspects of Coleoptera. Important works include *Classification and Biology* (1971) and *The Biology of the Coleoptera* (1981). He has been a member of the Coleopterist's Society since 1971, and was Vice-President 1971-1975.

Biographical Reference: [1979]. Darby, M. *A Biographical Dictionary of British Coleopterists*, pp. 133-134. Issued with *The Coleopterist's Newsletter*.

Collection

(1) One Ms notebook: *The Metendosternite in Coleoptera. Descriptive Notes and Sketches.*

102. CUNNINGTON, W.A.

Collection

(1) One Ms leaf: *Specimens Collected by W.A. Cunnington for Tanganyika Exploration Committee*, 1907.

Insect Room Lists: Volume 1, item 36.

103. D'ABRERA, Bernard (b. 1940)

A graduate of the University of New South Wales, D'Abrera collected butterflies in Europe, India, Sri Lanka, Malaysia, East Africa and throughout the Australasian region. A skilled photographer, he photographed specimens from The Natural History Museum, London and other collections for use in his huge publishing programme, *Butterflies of the World.*

Collection

(1) Annotated Ts: *Butterflies of the Oriental Region,* Pt 1. Papilionidae, Pieridae.

104. DALE, N. R.

Collection

(1) One letter: from N.R. Dale to J.W. Long, dated April 1823, concerning expenses for the funeral of a Mary Meggs. James Charles Dale (1792-1872), the nineteenth-century entomologist, is briefly mentioned in this letter. N.R. Dale is believed to be his father.

105. DARWIN, Charles Robert (1809-1882)

Born at Shrewsbury, he went to Shrewsbury School where he proved to be a poor student. Darwin was sent to Edinburgh University to study medicine but left in 1827 and prepared for holy orders at Christ College, Cambridge. Here he met John Steven Henslow, Professor of Botany, who stimulated his interest in natural history. Henslow recommended Darwin to accompany Capt. Robert Fitzroy on a five-year voyage in HMS *Beagle,* 1832-1836, to survey the coasts of Patagonia, Tierra del Fuego, Chile, Peru and various Pacific Islands. Darwin's work stemmed from observations and collections made during the voyage. The collections of animals made during the voyage were described in *The Zoology of the Voyage of* HMS *Beagle* (1839-1843). From his observations he developed a theory of evolution and provided evidence to explain how natural selection can produce adaptation in animals and plants. This revolutionary theory was published in the famous work *On the Origin of Species by Means of Natural Selection . . .* (1859). Other important publications include *The Variation of Animals and Plants under Domestication* (1868); *The Descent of Man and Selection in Relation to Sex* (1871) and *The Expression of the Emotions in Man and Animals* (1872).

Biographical References: 1882. Carrington, J.T. *The Entomologist* **15**:97-101; 1888. T.H.H. *Proceedings of the Royal Society, London* **1888**: i-xxv; 1991. Desmond, A. and Moore, J. *Darwin,* 808 pp. Michael Joseph, London.

Collection

(1) 26 Ms leaves: *Copy of Darwin's notes in reference to insects collected by him.* This provides a list of numbers referring to insects collected by Darwin during the voyage of the *Beagle* in the hand of Syms Covingdon, with additions and corrections in Darwin's hand.

Insect Room Lists: Volume 1, item 21.

106. DAVIES, William Maldwyn (1903-1937)

A worker in agricultural entomology for the Ministry of Agriculture. He had a personal interest in the Collembola and was working on a monograph of the British species at the time of his death.

Biographical Reference: 1937 Laing, F. *The Entomologist's Monthly Magazine* **73**: 92-93.

Collection

(1) Four Ms notebooks: [Notes and drawings of Collembola], c. 1935.

107. DEWITZ, Herman (1848-1890)

Dewitz was the Curator of the Natural History Museum, Berlin. During his life time he published a large number of papers in German entomological journals.

Biographical Reference: 1890. Mik, J. *Wiener Entomologische Zeitung* **9**:159-160, bibliography.

Collection

(1) Nine Ms leaves: [List of insects in various German collections], c. 1870s.

Insect Room Lists: Volume 1, item 17.

108. DE WORMS, Charles George Maurice, Baron (1903-1979)

Born in London, he was educated at Eton and King's College, Cambridge. Most of Charles De Worms' working life was spent as a chemist at the Institute of the Royal Cancer Hospital and Porton Down Experimental Station. He became interested in ornithology but the Lepidoptera were his main interest. He published a large number of papers and notes mainly in *The Entomologist* and *Entomologist's Record and Journal of Variation*. In 1962 he published *The Macrolepidoptera of Wiltshire*. De Worms was an extensive traveller, collecting Lepidoptera in Europe, North Africa, Australia, the West Indies and Canada. Most of his extensive collection was bequeathed to the Royal Scottish Museum in Edinburgh. He was a Fellow of the Royal Entomological Society of London and a member of both the London Natural History Society and the British Entomological and Natural History Society.

Biographical Reference: 1980. Chalmers-Hunt, J. M. *London Naturalist* **59**:88-89.

Collection

(1) Ms notes: [The Dioptidae].
(2) One Ms notebook: *Notes on the Genitalia of the Dioptidae. Genera* Dioptis *and* Josia.
(3) Two notebooks: *Genitalia of the Dioptidae Vols 1-2.*
(4) One notebook: *List of the species, subspecies and forms of the Dioptidae ex Seitz Vol.6.*
(5) Miscellaneous notes, lists and corespondence related to the Dioptidae.

109. DEYROLLE, Achille (1813-1865)

Born in Lille, he eventually settled in Brussels where he helped his father in the City Museum. Eventually he was offered the opportunity to go on a scientific mission to Brazil, and made considerable collections over the five months that he was there. During his lifetime he amassed a large collection of Coleoptera, but published very little.

Biographical Reference: 1866 Grenier, A. *Annales de la Société Entomologique de France* (4) **6**:132-134.

Collection

(1) One Ms : *Liste des Elaterides de Deyrolle Avril 1864.*

110. DISTANT, William Lucas (1845-1922)

His passion for natural history began to develop during a whaling voyage with his father in 1867. Much of Distant's early life was spent working in a London tannery, and whilst thus employed he made two long visits to the Transvaal. The journeys were related in *A Naturalist in the Transvaal* (1892). The second visit, of some four years, gave him time to amass a large collection of insects, many were described in *Insecta Transvaaliensia* (1900-1911). During the period 1899-1920 Distant was employed by The Natural History Museum, London, to work on Rhynchota. He became interested in exotic Rhopalocera, publishing *Rhopalocera Malayana* (1882-1886), but he devoted the major part of his time to the Rhynchota.

Biographical Reference: 1922. Anon. *The Entomologist's Monthly Magazine* **58**:66-67.

Collection

(1) Loose-leaf Ms: [*Platypleura beiraca*] unpublished.
(2) Loose-leaf Ms: [*Acanthapsis aurantiaca*] unpublished.

111. DOLLMAN, Hereward Chune (1888-1919)

In a tragically short career Dollman achieved a considerable amount of work. His interest in Coleoptera started at an early age and at 16 he was elected a Fellow of the Entomological Society of London. After university he went to Africa to work on tsetse flies and to carry out a survey on sleeping sickness. He was a talented artist and whilst in Africa drew a great many lepidopterous larvae, particularly of the Sphingidae, for a proposed publication. This was never completed. All his collections came to The Natural History

Museum, London on his death.

Biographical Reference: 1919. J.G.D. *The Entomologist's Monthly Magazine* **55**:39-140.

Collection

(1) Two letters: [Concerning the offer of the Dollman collection to the Natural History Museum], 1919.
(2) Five collecting diaries: 1909-1912.
(3) One Ms notebook: *Coleopterous fauna of Ditchling, Sussex and the surrounding area*, c. 1912.
(4) Two collecting notebooks: *Coleopterous fauna of Sussex.*
(5) Three Ms notebooks: *Coleoptera taken personally*, c. 1908-1912.
(6) One Ms notebook: *Botanical notebook*; contains drawings.
(7) One Ms notebook: *British Coleoptera data.*
(8) One Ms notebook: *Phytophagous Coleoptera and their food plants.*
(9) One Ms notebook: *Notes relative to plates of Lepidoptera larvae*, 1913.
(10) One notebook: [List of British plant and insect specimens and localities].
(11) One folio volume: *Drawings of Larvae; North West Rhodesian Lepidoptera with notes and data.* Includes 166 drawings, c. 1914-1915.
(12) One Ms volume: *Data of Lepidoptera in the Cabinets of H.C. Dollman 1916-17-18 North West Rhodesia.*
(13) Two loose-leaf Mss: [Notes relating to the Dollman drawings of Rhodesian lepidoptera].
(14) One loose-leaf Ms: *Notes on Dollman's Drawings of Rhodesian Lepidoptera* by N.D. Riley.
(15) One bound volume: [Copies of Dollman's Drawings of Lepidopterous larvae] by F.W. Frohawk.

112. DONCKIER, Henri

Collection

(1) One Ms leaf:[List of specimens sold to the British Museum], 1904.

Insect Room Lists: Volume 2, item 61.

113. DONISTHORPE, Horace St John Kelly (1870-1951)

Donisthorpe began to study medicine at Heidelburg, but gave up the idea of this profession. From about 1890 he started collecting beetles; this became his biding passion, only being content with specimens that he caught himself, and not those given and collected by others. His study of beetles associated with ants led him on to the study of British ants. He published two books, *British Ants* (1915) and later *The Guests of British Ants* (1927). These are still well regarded as standard works on the subject. His collection of beetles and ants, and his natural history library were acquired by The Natural History Museum, London. He was a Fellow of the Royal Entomological Society and one time Secretary and senior member of the Entomological Club.

Biographical Reference: 1951. R.B.B[enson]. *Journal of the Society for British Entomology* **4**:23-24.

Collection

(1) Five files: [Notes on the distribution of ants in the British Isles].
(2) One scrapbook: Cuttings from the Proceedings of the Entomological Society of London, 1891.
(3) One scrapbook: contains copies of reviews and letters relating to *British Ants* (1915).
(4) Five Ms volumes: *Journal of British Coleoptera captured by H. Donisthorpe,* 1879-1883; 1902-1914; 1915-1930; 1930-1938; 1938-1940. Inscription reads: *A list of the British Coleoptera taken by me and in my collection with dates, localities and notes on how captured, arranged in the form of a journal.*
(5) Six collecting notebooks: [Localities and food plants of Hymenoptera].
(6) One Ms notebook: [List of Coleoptera and Hymenoptera duplicate specimens including the names of people to whom the specimens were given].
(7) One Ms notebook: [List of the Coleoptera of Cambridgeshire].
(8) Annotated copy of *Catalogue of British Coleoptera* (1904).
(9) Correspondence collection: approximately 200 letters to Donisthorpe, 1900-1948; includes letters from:

Allen, A.A.	1940-1948
Beare, Thomas Hudson (1859-1940)	1900-1936
Creighton, William Steel (1902-1973)	1937
Eltringham, Harry (1873-1941)	1937
Green, Edward Ernest (1861-1949)	1911-1933
Keys, James H G (1855-1941)	1907-1935
Smith, Marion Russel	1938-1939

114. DONOVAN, Lieut.-Col. Charles (1863-1951)

Born in Calcutta, Donovan was educated at Queen's College, Cork, and later Trinity College, Dublin. In 1891 he was commissioned into the Indian Medical Service, and finally settled at the Government General Hospital in Madras, returning to England on his retirement in 1919 with the rank of Lieutenant-Colonel. Originally interested in birds, he turned his attention to Lepidoptera after his arrival in Madras, and after his retirement concentrated completely on entomology. In 1936 he published a *Catalogue of the Macrolepidoptera of Ireland*. His collections were acquired by The Natural History Museum, London.

Biographical Reference: 1952. Riley, N. D. *The Entomologist* **85**:120.

Collection

(1) One Ts volume: *Butterflies of Southern India including Mysore and Travancore*, includes manuscript annotations by Donovan.

115. DOUBLEDAY, Edward (1811-1849)

Doubleday developed an early interest in natural history and travelled to North America, where he collected many insect specimens. On his return he worked at the British Museum, where he studied the Lepidoptera collections. His publications include *List of the Specimens of Lepidopterous Insects in the British Museum* (1844-1848) and, with J. O. Westwood, *The Genera of Diurnal Lepidoptera* (1846-1852).

Biographical Reference: 1850. Douglas J.E. *Zoologist* **8**:2712.

Collection

(1) Eight Ms volumes: believed to be for *List of the Specimens of Lepidopterous Insects in the British Museum* (1844-1848).

116. DOUBLEDAY, Henry (1809-1875)

A careful, conscientious observer. His foremost interest was in the British Lepidoptera. He was not a prolific writer but had an extensive correspondence and was liberal with his supply of specimens to those interested in the subject. He was also a general naturalist with a particular interest in ornithology.

Biographical Reference: 1978. Mays, R. *Henry Doubleday: The Epping Naturalist.* 118pp. Marlow: Precision Press.

Collection

(1) Two Ms leaves: *Lepidoptera from Madeira.* An added Ms note reads: *T.V. Wollaston Collection: Madeira B.M.1858:21. Determinations of Lepidoptera of Madeira in handwriting of Henry Doubleday this list was in the last drawer of the Madeiran Cabinet.*

Insect Room Lists: Volume 2, item 109.

(2) A framed oil painting of Doubleday hangs in the Entomology Library.
(3) One autograph letter from Doubleday to Thomas Beck, 1829.
(4) One autograph letter from Doubleday, 1875.

117. DOUGLAS, John William (1814-1905)

A one time President of the Royal Entomological Society and editor of *The Entomologist's Monthly Magazine,* he became interested in insects whilst working at Kew. He published many papers and a number of books. Although his main interest was the Lepidoptera, Douglas was joint author of the work *The British Hemiptera Vol.1. Hemiptera-Heteroptera* (1865). He was a social man who was never happier than among his fellow entomologists; he was a great encourager of the younger entrants to the subject.

Biographical Reference: 1905. E.S. *The Entomologist's Monthly Magazine* **41**:221-222.

Collection

(1) Three Ms notebooks:
- (i) *Captures 1 1848-52.*
- (ii) *Captures 2 1853-56.*
- (iii) [Captures] 3 *1855-96.*

(2) One Ms leaf: [List of the Douglas insect collection].

Insect Room Lists: Volume 2, item 85(2).

118. DRIESCHE, A. Van

Collection

(1) One Ms leaf: *Eggs of locusts* 1907.

Insect Room Lists: Volume 1, item 9.

119. DRURY, Dru (1725-1803)

Born in London, the son of a silversmith. By virtue of his marriage and inheritance of the family business he was a reasonably wealthy man who could afford to support his most serious hobby of entomology. Drury's collection had great fame during his life time. He spent much time and money persuading others to collect specimens for him from foreign countries. He had a wide correspondence with entomologists around the world. Linnaeus (1707-1778) and William Kirby (1759-1850) both named species after him. Between 1770 and 1782 he published his three-volume work, *Illustrations of Natural History*, with the majority of the plates drawn and engraved by Moses Harris (1731-1788). His *Directions for Collecting Insects in Foreign Parts* was published in 1800 and translated into several languages.

Biographical Reference: 1927. Weiss, H.B. *Entomological News* **38**:208-214.

Collection

(1) Photocopy of Ms: *A catalogue of the exotic insects in the collection of Dru Drury,* 1784. From the original held in Hope Entomological Collections, Oxford. It was from these lists that E. Donovan prepared the Drury Sale Catalogue.

(2) One Ms notebook: [Giving an account of the printing expenses incurred by the publication of *Illustrations of Natural History*], 1770-1782.

(3) Two Ms diaries:

- (i) Vol. 1 covers the period July 1794 to April 1796. It includes general notes such as household tips, accounts, medical notes and wine production. It also includes notes on the coloration of plates by Moses Harris in the work *Illustrations of Natural History* (1770-1782).
- (ii) Vol. 2 covers the period 1801-1805. Includes notes on the *Illustrations of Natural History* (1770-1782).

(4) Six Ms leaves: a geneological table of the Drury family, c. 1800.

(5) One volume of miscellaneous papers, includes the following items.

- (i) Ms: *Account of boxes with instruments for catching insects delivered to different persons* 1758.
- (ii) Mss: [Lists of boxes delivered onto ships between 1768-1771].
- (iii) [Agreements between Drury and ships captains regarding gold prospecting].
- (iv) [Statements – listing the Drury estates].
- (v) Ms: *An account of the contents of Mr Lewin's Chests*, 1790.
- (vi) Ms: *Invoice of goods delivered on board the Buffalo for Mr J.W. Lewin, the return to be made in Insects from Port Jackson* [Australia], 1798.
- (vii) Ms: *An account of the contents of different chests of boxes sent [to] Mr. Lewin.*
- (viii) Ms: *Catalogue de la collection des Insectes de Dru Drury de Londres* 1780.
- (ix) Correspondence collection, including letters from:

Graves, Charles	1802
Hill, William	1803
Jarratt, Devereaux	1769
Lewin, John William (1710-1819)	1800-1803
Linnaeus, Carl (the younger)	1780
Smeathman, Henry (1750-1787)	1778

(6) One folio letterbook: contains copies of letters made by Drury to correspondents between 1761 and 1783. A complete listing of these letters is given by C.D. Sherborn, 1937. *Journal of the Society for the Bibliography of Natural History* **1**:109-111. Copies include letters to:

Abbot, J. (1751-1840)	1774
Bentinck, Margaret Cavendish (Dowager Duchess of Portland) (d. 1785)	1773-1775
Brunnich, Martin Throne (1737-1827)	1770-1772
Harris, Moses (1731-1788)	1770
Hunter, William (1718-1783)	1768-1769
Linnaeus, Carl von (1707-1778)	1770
Pallas, Peter Simon (1741-1811)	1764-1811
Smeathman, Henry (1750-1787)	1768-1775

120. DUBOIS, Charles (1656-1740)

Remembered now as a botanist and horticulturalist, he was a Fellow of the Royal Society and had wide interests, including entomology. He amassed a large collection of botanical specimens and contributed plants to the Sloane Herbarium.

Biographical Reference: 1989. Jessop, L. *Bulletin of the British Museum (Natural History)* Historical Series **17**:1-165.

Collection

(1) One Ms notebook: *Ms notes on insects with original drawings written by a collector in 1692 & 1695.* Includes notes and observations on insects caught in and around Mitcham, South London. Another later manuscript in another

hand is also included: *Miscellany thoughts about vegetables.*

121. DURRANT, John Hartley (1863-1928)

Durrant was born at Hitchin in Hertfordshire, the son of a schoolmaster. He developed an interest in Microlepidoptera, and in 1885 Thomas de Grey, afterwards Lord Walsingham (1843-1919), appointed him as his private secretary and entomological assistant. He had charge of Walsingham's collections and helped build his library. In 1910 Walsingham donated his collections and library to The Natural History Museum, London, B.M.1910-427; he also gave sufficient funds to pay the salary of Durrant, who was to continue to look after the collection on its transfer. Whilst on the Museum staff Durrant headed an inquiry into the army biscuit. In 1913 he published, a preliminary report of the temperature reached in Army biscuits during baking, especially with reference to the destruction of the imported flour moth (*Ephestia kuhniella* (Z) in the *Journal of the Royal Army Medical Corps.* **20**:615-634, plates 1-7. Durrant served on the Library Committee of the Entomological Society of London and was the first secretary of the British National Committee on Entomological Nomenclature.

Biographical Reference: 1928. Riley, N.D. *The Entomologist* **61**:73-75, portrait.

Collection

(1) Three Ms notebooks and one loose-leaf Ms relating to Jacob Hübner (1761-1826) specimens.

122. DYAR, Harrison Gray (1866-1929)

Born in New York City, Dyar graduated from the Massachusetts Institute of Technology and in 1889 obtained a Ph.D. at Columbia University. He was a wealthy man and from 1897 until his death he worked without pay as an Honorary Custodian of the Lepidoptera at the US National Museum, Washington D.C. He became an authority on the Culicidae of the Americas. He collaborated with Frederick Knabb and L.O. Howard in writing the four-volume work, *The Mosquitoes of North and Central America and the West Indies*

(1912-17). Dyar was editor of the *Journal of the New York Entomological Society*, 1904-1907; the *Proceedings of the Entomological Society of Washington*, 1909-1912 and *Insecutor Inscitiae Menstruus*, 1913-1926.

Biographical Reference: 1974. Knight, K.L. *Mosquito Systematics* **6**:11-24, bibliography.

Collection

(1) Four Ms leaves: *Lists of revised and returned insects by H.G. Dyar*, 1906. Includes insects from Kamchatka presented by G.E.H. Barrett-Hamilton.

Insect Room Lists: Volume 1, item 30.

123. EASTON, Alan Maurice (1907-1989)

Easton studied medicine at St Thomas's Hospital, London and later opened a medical practice in Great Bookham, Surrey where he worked for 52 years. As a police surgeon he developed an interest in forensic entomology. However, his great entomological interest concerned the genus *Meligethes* (Coleoptera). He carried out much pioneering work on these beetles; his collection was donated to The Natural History Museum, London.

Biographical Reference: 1990. Bacchus, M. and Kirk-Spriggs, A. *Antenna* **14**:61-63.

Collection

(1) Correspondence collection: approximately 500 letters sent from various British entomologists to Easton c. 1947-1978 concerning his work on *Meligethes*. Correspondents include:

Allen, A.A.	1947-1964
Buck, F.D.	1949-1963
Coulson, F.J.	1947
Frey, G.	1950-1978
Hesse, A.J.	1966-1978

Janestscher, H.	1967
Johnson, C.	1962-1968
Last, H.	1946-1968
Lewis, E.	1947-1964
Nakane, T.	1955-1964
Osborne, P.	1960-1967
Spornraft, K.	1964-1969

(2) Approximately 100 drawings of *Meligethes.*

124. EATON, Alfred Edwin (1845-1929)

Interested in many aspects of entomology, especially Diptera, Eaton is best remembered for the work A Monograph of the Ephemeridae, published in *Transactions of the Entomological Society of London* (1871), 1-164. In 1875 Eaton joined an expedition to Kerguelen Island where he discovered many interesting insects. He later studied the Psychodidae and collected specimens from Algeria. Over a period of years he presented many specimens from Britain, Switzerland and Algeria to The Natural History Museum, London.

Biographical Reference: 1929. Edwards, F.W. *The Entomologist* **62**: 167-168.

Collection

(1) One Ts: *An unscientific analytical synopsis of the Genera of British Ephemeridae,* c. 1900 (This synopsis was prepared by Eaton for F.M. Halford. It includes annotations by Halford, K.G. Blair and M.E. Mosely.)

125. EATON, Mary Emily (1873-1961)

Collection

(1) 30 watercolour drawings: Jamaican Lepidoptera, 1909-1910, with manuscript notes.

126. EDELSTEN, Hubert McDonald (1877-1959)

Edelsten was a keen Lepidopterist and contributed hundreds of notes and articles to *The Entomologist*. During the First World War he served in the Royal Naval Air Ambulance Service. After the War he became Secretary to the Committee for the Protection of British Insects. In 1939 he joined the Ministry of Agriculture and was concerned with the entomology of food crops. This work included the study of the Colorado Beetle. Publications included an account of the Lepidoptera of Cambridgeshire (with John Fryer) in the *Victoria County History of Cambridgeshire and the Isle of Ely*, Vol.1. (1938) and a revision of Richard South's *The Moths of the British Isles* (1939). Edelsten was an Honorary Fellow of the Royal Entomological Society of London and an Associate in charge of British Lepidoptera at The Natural History Museum, London. His collection of 780 Lepidoptera specimens and 208 Odonata specimens was presented by his daughter to The Natural History Museum, London. B.M.1959-304.

Biographical Reference: 1959. Riley, N.D. *The Entomologist* **92**:155-157, portrait.

Collection

(1) Loose-leaf Ms: *Contributions to the Life history of* Oria musculosa, c. 1944.

127. EDWARDS, Frederick Wallace (1888-1940)

Born at Fletton, Peterborough he entered Christ College Cambridge in 1906. Four years later he was appointed to the staff of the Entomological Section of the Zoology Department of The Natural History Museum, London, working on the Nematocerous Diptera and in particular the Culicidae. He made two major expeditions. In 1926 he travelled to Argentina and Chile, bringing back over 20,000 specimens including 700 new species, and in 1934 he collected in East Africa. He published over 350 papers, describing 2,000 new species. These included major contributions to the *Diptera of Patagonia and South Chile* (1929-1951). He was a co-author of *British Blood Sucking Flies* (1939) with H. Oldroyd and J. Smart. Edwards was elected a Fellow of the Royal Society in 1938.

Biographical References: 1941. Smart, J. *The Entomologist* **74**:22-24, portrait; 1945. Smart, J. *Journal of the Society for the Bibliography of Natural History* 2:19-34, bibliography.

Collection

(1) Ms collection: Includes:
- (i) *List of Chironomidae in the Cambridge University Museum of Zoology*
- (ii) *East Indian Dipt*[era]. *Nem*[atocera] *in Amsterdam Museum.*
- (iii) [A.E.] *Holmgren's Spitzbergen Nematocera & Bear Island. Notes made in Stockholm Museum,* 1923.
- (iv) *[J.W.] Meigen's* Sciara *in Paris Museum.*
- (v) *Notes on Meunier's types of amber Mycetophilidae,* 1933.
- (vi) Miscellaneous notes and drawings.

(2) Memorandum from Edwards to C. J. Gahan regarding Edwards' style of dress, 1915.

128. ELIOT, John Nevill

Eliot's father was a keen lepidopterist and passed his enthusiasm on to his son. J. N. Eliot was a professional soldier, eventually being promoted to the rank of Colonel. He has published many entomological papers and edited revised editions of *The Butterflies of the Malay Peninsula* (1978 and 1992) by A. S. Corbet and H. M. Pendlebury.

Collection

(1) One Ms notebook: *Colonel Eliot's observations on the Riviera at Cavalaise,* 1949.
(2) Three Ms notebooks: *Observations for dates of Lepidoptera round St Jean-de Luz,* c. 1947-1960.
(3) 12 Ms notebooks: Lepidoptera notes and observations, c. 1947-1955.

129. ELLIOTT, Ernest. A. (d. 1936)

A keen hymenopterist, he was a Fellow of the Zoological Society and Royal Entomological Society of London.

Biographical Reference: 1936. Imms, A.D. *Proceedings of the Royal Entomological Society London* (C) **1**:55.

Collection

(1) One bound Ts: *Monograph of the Hymenopterous Family Stephanidae,* 1920. (This was published in the *Proceedings of the Zoological Society of London 1922,* pp. 705-834.)

130. EMDEN, Fritz Isidore van (1898-1958)

Born in Amsterdam, van Emden studied natural history at the University of Leipzig. He became a professional entomologist working at the Deutsches Entomologisches Institut in Berlin and the Natural History Museum in Dresden. In 1936 he moved to England, joining the staff of the Imperial Institute of Entomology as a dipterist, although his personal interest was always the Coleoptera. His publications include the Tachinidae and Calliphoridae volume in the series, *Handbooks for the Identification of British Insects,* Vol. X. Pt.4(a) (1954), published by the Royal Entomological Society of London, and *Diptera* Vol. 7 (1965) on the Muscidae, in the *Fauna of India.*

Biographical Reference: 1958. Oldroyd, H. *The Entomologist's Monthly Magazine.* **94**:228-229, portrait.

Collection

(1) One loose-leaf Ms: [Coleoptera especially the Rhipiceridae].

131. ENTOMOLOGICAL MISCELLANEA

Collection

(1) A miscellaneous collection of nineteenth-century entomological items including advertisements, newspaper cuttings, entomological society leaflets. Also includes:

- (i) One Ms: *List of Native Lepidoptera collected during the summer of 1857,* by the Master Mason of Little Friday Hill, Chingford, Essex.

132. EVANS, R. du B.

Collection

(1) One Ms notebook: *Notes on Butterflies and Moths,* c. 1924.

(2) One Ms notebook: *Butterflies and Moths*. Relates to collection from South America donated to The Natural History Museum, London. B.M.1934-29.

133. EVANS, William Harry (1876-1956)

Born in Shillong, Assam, Evans was sent as a boy to King's School, Canterbury. When he was 18 years old he joined the Royal Engineers serving in India, Somaliland and France, eventually attaining the rank of Brigadier. Throughout his service in India he collected butterflies. His *Keys for the Identification of Indian Butterflies* was originally published in parts in the *Journal of the Bombay Natural History Society* and was later produced as a monograph in 1927. On his retirement in 1932 he settled close to The Natural History Museum in London. There he made a particular study of the Hesperiidae, culminating in a series of catalogues of the *Hesperiidae in the British Museum (Natural History)*, published in three series: *Africa* (1937); *Europe, Asia and Australia* (1949) and *America* (1951-1955). He was a Fellow of the Royal Entomological Society of London.

Biographical Reference: 1956. Riley, N.D. *Lepidopterist's News* **10**:193-199

Collection

(1) Loose-leaf Ms: *A Catalogue of the American Hesperiidae in the British Museum (Natural History)*, Pts 1-4, c. 1950.

(2) Nine plates of original drawings: *Catalogue of American Hesperiidae,* Pt.1, c. 1950.

(3) Loose-leaf Ms: *A Catalogue of the Hesperiidae in the British Museum (Natural History) from Europe, Asia and Australia*, c. 1949.

(4) One Ms notebook: *Indian Butterflies Lycaenidae.* Contains nature prints of specimens.

(5) One album: *Indian Butterflies.* Contains nature prints of specimens.

(6) Two Ms notebooks: *A Review of the* Amblypodia *Group (Lepidoptera - Lycaenidae).*

(7) Three Ms notebooks: Addenda and corrigenda to the *Catalogue of Hesperiidae in the British Museum.*

(8) Two Ms notebooks: *'Arhopala'*.

(9) One Ms collecting diary: Localities St Vincent, W. Indies, c. 1889.

(10) Two Ms notebooks: containing nature prints and notes.

(11) One Ms notebook: *Catalogue of Coleoptera from Grenada*, c. 1891.

(12) One MS loose-leaf notebook: *Register of B.M. Types of Hesperiidae.*

(13) Original drawings : *Catalogue of the African Hesperiidae*, 1937.

(14) Original drawings: *Catalogue of the Hesperiidae from Europe, Asia and Australia*, 1949.

134. FASSNIDGE, William (1888-1949)

For most of his working life Fassnidge was employed as a modern language master in Southampton. His particular interest was collecting Palaearctic Lepidoptera and the varieties of *Peronea cristana*. He was a member of the South London Entomological and Natural History Society and a Fellow of the Royal Entomological Society of London.

Biographical Reference: 1949, S.N.A.J. *The Entomologist's Record and Journal of Variation* **61**:58-59.

Collection

(1) Correspondence collection: approximately 100 miscellaneous letters sent to Fassnidge and others, c. 1926.

135. FELDER, Cajetan von (1814-1894)

In 1848 Felder became a barrister and during his life stood three times for election as Mayor of Vienna. With the help of his son Rudolf Felder (1842-1871) he amassed a huge entomological collection. His publications culminated with his contributions to the work, *Reise der Österreichischen Fregatte Novara 1857-59*, Zoologischer Theil, Lepidoptera, Rhopalocera (1864-

1867) by C. and R. Felder. His collection of Lepidoptera and Coleoptera were sold to Lionel Walter Rothschild (1868-1937), who in turn bequeathed his collection to The Natural History Museum, London in 1937.

Biographical Reference: 1894. Rogenhofer, A.F. *Deutsche Entomologische Zeitschrift, Iris* **7**:363.

Collection

(1) Approximately 1,000 letters sent to Felder, c. 1856-1891. Includes letters from:

Bates, Henry Walter (1825-1892)	1859-1869
Becker, Alexander (1818-1901)	1857
Becker, Theodor (1840-1928)	1856-1859
Behr, Hans Herman (1818-1904)	1870-1871
Bilimek, Dominik	1869-1871
Boheman, Carl Heinrich (1796-1868)	1859-1863
Boisduval, Jean Baptiste Alphonse Déchauffour de (1799-1879)	1872
Bremer, Otto (d. 1873)	1859-1869
Brunner von Wattenwyl, Carl (1823-1914)	1870
Butler, Arthur Gardiner (1844-1925)	1869-1870
Castelneau, François Louis Nompar de-Caumont de Laporte de (1810-1880)	1860-1868
Chaudoir, Maximilien de (1816-1881)	1860-1868
Dohrn, Carl August (1806-1892)	1856-1860
Doleschall, Carl Ludwig (1827-1859)	1856-1859
Edwards, Henry (1830-1891)	1870-1876
Edwards, William Henry (1822-1909)	1867-1876
Ferrari, Johann Angelo (1806-1876)	1869
Fitzinger, Leopold Joseph (1802-1884)	1865
Frauenfeld, Georg Ritter von (1807-1873)	1867
Herrich-Schäffer, Gottlieb August W. (1799-1874)	1859-1864
Hewitson, William Chapman (1806-1878)	1878
Higgins, Edmund Thomas	1869-1871

Kheil, Napoleon Manuel (1849-1923)	1888-1891
Lederer, Julius (1821-1870)	1870
Maasen, J. Peter (1810-1890)	1878
Ménétriés, Edouard (1802-1861)	1858-1861
Mniszech, Georges Vandalin (d. 1881)	1869
Moore, Frederic (1830-1907)	1881
Nicéville, Charles Lionel Augustus de (1852-1890)	1890
Niepelt, Wilhelm (1862-1936)	1887
Oberthür, Charles (1845-1924)	1870
Ribbe, Heinrich (1832-1898)	1891
Sallé, Auguste (1820-1896)	1860-1862
Salvin, Osbert (1835-1898)	1864-1871
Saussure, Henri Louis Frederic de (1829-1905)	1866
Schaufuss, Ludwig Wilhelm (1833-1890)	1861
Scott, Alexander Walter (1800-1883)	1873
Sikora, Franz (d. 1902)	1890
Snellen van Vollenhoven, Samuel Constant (1816-1880)	1860-1873
Staudinger, Otto (1830-1900)	1872-1890
Stevens, Samuel (1817-1899	1860-1870
Stichel, Hans (1862-1936)	1890-1891
Stoliczka, Ferdinand (1838-1874)	1864-1872
Suffert, Ernst (d. 1907)	1866-1869
Thiele, Hermann (1841-1918)	1890
Trimen, Roland (1840-1916)	1867-1868
Walker, Francis (1809-1874)	1869-1870
Wallace, Alfred Russel (1823-1913)	1862-1869
Wallengren, Hans Daniel Johan (1823-1894)	1869
Watkins, William (1849-1900)	1895
Westwood, John Obadiah (1805-1893)	1871

136. FENN, Charles (1840-1925)

For most of his working life he was employed in the Bank of England. Fenn was a keen lepidopterist and discovered *Nonagria brevilinea* in the Norfolk fenlands.

Biographical Reference: 1926. Adkin, R. *The Entomologist* **59**:120

Collection

(1) Two Ms volumes : *Lepidoptera Data*. Contains an alphabetical list of captures and their localities, c. 1860-1893.
(2) One Ms volume: *Entomological Notes etc. with Larval Index.*
(3) One Ms volume: *Descriptions of Lepidopterous Larvae.*

137. FENYES, Adelbert (1863-1937)

Born in Hungary, he trained as a doctor and practised medicine for several years in Egypt. Later Fenyes moved to California and developed a keen interest in ornithology and entomology. He collected in America and Mexico and amassed an extensive collection which was purchased by the California Academy of Sciences. He became a world authority on the Staphylinid tribe Aleocharinae.

Biographical Reference: 1937. Fall, H.C. *Pan-Pacific Entomologist* **13**:145-147, portrait.

Collection

(1) One Ms leaf: *Collection of N. Am[erican] Aleocharinae from Dr. A. Fenyes.* B.M.1910-402.

Insect Room Lists: Volume 2, item 89.

138. FLETCHER, Josh

Collection

(1) One Ms notebook : *A Catalogue of Bruzilian Insects in the Collection of Josh Fletcher Esq*, 1829.

139. FLETCHER, Thomas Bainbrigge (1878-1950)

A one time naval paymaster, in 1910 he transferred from the navy to be Imperial Entomologist to the Government of India. He had interests in all spheres of entomology, including the problems of economic pests in India. His prime concern, however, was the Microlepidoptera on which he concentrated after his retirement. During World War II Bainbrigge Fletcher was of considerable help to The Natural History Museum, London, allowing part of its valuable insect collections and libraries to be stored at his home in the Gloucestershire countryside. Among his major publications are *Some South Indian Insects* (1914) and *A List of Generic Names used for Microlepidoptera* (1929).

Biographical Reference: 1952. Sen, S.K. *Indian Journal of Entomology* **14**:87-90, portrait.

Collection

(1) One Ms notebook: *List of Captures*, 1880-1920. Specimens now in The Natural History Museum, London.
(2) Five private journals: entomological notes written whilst serving on various naval ships:
  (i) HMS *Emanuel* and HMS *Centurion*, in Korea and Singapore, 1895-1896.
  (ii) HMS *Centurion* in Vladivostok and Yokohama, 1899.
  (iii) HMS *Edgar* in Colombo and Suez Canal.
  HMS *Gladiator* in Morocco, 1900-1901.
  (iv) HMS *Sealark* at various ports, 1908.
(3) One Ms notebook: List of captures whilst serving on HMS *Centurion* in China, 1899.
(4) One letterbook: 1910-1913 (a few additions 1944-1945).
(5) 22 diaries: mostly collecting data, 1921-1947.
(6) Three boxes of Ms notes: [Notes on *Oxyptilus, Agdistis and Platyptilia*].
(7) Seven folders of Mss: [Notes on Microleidoptera families].
(8) Correspondence collection containing approximately 800 letters, including letters from:

Audcent, Henri, L.F. (1875-1951)	1943
Beirne, Bryan P. (b. 1918)	1939-1949
Burr, Malcolm (1878-1954)	1910-1950
Caradja, Aristide (1861-1955)	1901-1939
Clutterbuck, Charles Granville (1871-1957)	1935-1949
Cockayne, Edward Alfred (1880-1956)	1923-1948
Curtis, William Parkinson (1878-1968)	1933-1946
Dannreuther, T. (1873-1963)	1935-1948
Diakonoff, Alexander Mikhailovich (1907-1989)	1938-1947
Ford, Leonard Talman (1880-1961)	1935-1948
Green, Edward Ernest (1861-1949)	1906-1938
Hallet, Howard Mountjoy (1878-1958)	1942-1945
Hayward, Kenneth J. (1891-1972)	1937-1938
Hering, Erich Martin (1893-1967)	1933-1938
Janse, Antonie Johannes Theodorus (1877-1970)	1935-1944
Jones, Frank Morton (1869-1962)	1925-1935
Jordan, Heinrich Ernst Karl (1861-1959)	1936-1940
Lhomme, Léon (1867-1949)	1934-1936
Marshall, Guy Anstruther Knox (1871-1959)	1934-1938
Marshall, John Frederick (1874-1959)	1936-1945
Obraztsov, Nikolaus Sergeyevich (1905-1966)	1948-1949
Strand, Embrik (1876-1947)	1922-1936

140. FOSTER, W.

Collection

(1) Four Ms leaves: *List of the Coleoptera collected by the Sladen Expedition to Santa Anna da Chapada in Matto Grosso, and Central Brazil*, 1903.
(2) Six Ms leaves: *Collections sent for determination by W. Foster from Sapuray, Paraguay*, 1903.

Insect Room Lists: Volume 1, items 19,20.

141. FOUNTAINE, Margaret Elizabeth (1862-1940)

Born near Norwich, she developed an early interest in butterflies. Margaret Fountaine travelled through Europe, Africa, India, America and the West Indies collecting specimens. Her collection of over 22,000 butterflies was bequeathed to the Castle Museum, Norwich.

Biographical Reference: 1980. Cater,. W. F. (ed.) *Love Among the Butterflies: The Travels and Adventures of a Victorian Lady.* 223pp. Collins, London.

Collection

(1) Four volumes: 862 original watercolour drawings of Lepidoptera larvae and pupae with food plants.

- (i) Vol.1 *Larvae and Pupae from South America, India, Ceylon and Malaya, 1907-1914.*
- (ii) Vol.2 *Larvae and Pupae from North America, West Indies, Central America, Malaya, Java, Australia, Fiji Islands, New Zealand, France, Burma, China, Philippines, Canary islands and West Africa, 1910-1926.*
- (iii) Vol. 3. *Larvae and Pupae from West Africa, Canary Islands, South America and United States, 1926-1931.*
- (iv) Vol.4. *Larvae and Pupae, 1931-1939.*

(2) One folder: Correspondence and notes regarding the probate of the Fountaine will, between her executors and N.D. Riley of The Natural History Museum, London, 1940-41.

(3) One folder: Correspondence between N.D. Riley and M. Fountaine, 1932-1939.

142. FRASER, Frederic Charles (1880-1963)

Born at Woolwich, he developed an early interest in natural history and in particular the Odonata. Fraser studied medicine and in 1907 joined the Indian Medical Service. He had a number of apppointments in India and made extensive collections of dragonflies from India, Ceylon and Burma. He published over 300 papers and publications, including three volumes on Odonata in the *Fauna of British India* (1933-1936), *A Reclassification of the order*

Odonata (1957) and *A Handbook of Dragonflies of Australasia* (1960).

He described hundreds of new species and presented type material to The Natural History Museum, London. He was a Fellow of the Royal Entomological Society of London and of the Royal Zoological Society of New South Wales.

Biographical Reference: 1963. Kimmins, D. *The Entomologist* **96**:94-95.

Collection

(1) Correspondence: 70 letters from F.C. Fraser to A.E. Gardner and 20 letters from Gardner to Fraser, c. 1949-1958.
(2) One Ms notebook: [Notes on Orthoptera and Odonata wings].
(3) One Ts and drawings: [A proposed revision of Lestidae] (incomplete).
(4) One volume: *Sketches from living Odonata or from spirit specimens. Drawn over a period of twenty years,* c. 1920-1940.
(5) One volume of original watercolour drawings *Lepidoptera Indica larvae,* 1908-1918.

143. FREY, Heinrich (1822-1890)

Born in Frankfurt, his interest in entomology, particularly the Lepidoptera, developed from an early age. In 1849 he settled in Zurich, accepting a professorship in the medical faculty at the University in 1851. It was in Zurich that he met his lifelong friend Jacob Boll, who contributed much to Frey's first major work concerning the microlepidoptera, *Die Tineen und Pterophoren der Schweiz,* published in 1856. Besides publishing many entomological papers, Frey also had a considerable reputation as a medical writer.

Biographical Reference: 1890. Stainton, H.T. *The Entomologist's Monthly Magazine* **26**:113-117.

Collection

(1) Approximately 200 letters to Frey, mostly from European entomologists c. 1856-1885. Includes letters from:

Boll, Jacob (1828-1880)	1871-1876
Herrich-Schäffer, Gottlieb August Wilhelm (1799-1874)	1851-1857
Ragonot, Emile Louis (1843-1895)	1879-1882
Rougemont, Fréderic (1838-1917)	1882-1886
Stainton, Henry Tibbats (1822-1892)	1855-1883
Turati, Gianfranco (1861-1905)	1881-1855
Zeller, Philip Christoph (1808-1883)	1855-1882

144. FROHAWK, Frederick William (1861-1946)

Born at Brisley Hall, Norfolk, he developed an early love of natural history, and displayed a natural talent for painting and drawing. During his life he was commissioned to illustrate many publications including *The Field*, and many natural history books. Lord Rothschild employed him as an artist, as did The Natural History Museum, London. His particular entomological interests centred on British butterflies. For over twenty years he bred every species of British butterfly, and published his observations and illustrations in the classic work *The Natural History of the British Butterflies* (1924). In 1934 he published *The Complete Book of British Butterflies*, which was followed in 1938 by the *Varieties of British Butterflies*. He was a Fellow of the Entomological Society of London and became a Special Life Fellow in 1926.

Biographical Reference: 1987. Chatfield, J. *F.W. Frohawk His Life and Work.* 184pp. Crowood Press, London.

Collection

(1) Seven boxes : manuscripts, drawings, proofplates and notebooks relating to *The Natural History of the British Butterflies* (1924).
(2) Two volumes of original watercolour drawings: *The Natural History of British Butterflies.*
(3) One volume of 48 original watercolour drawings, *Varieties of British Butterflies.*
(4) 40 original watercolour drawings: [British Lepidoptera]. Published by The Natural History Museum as a series of postcards.

145. GABRIEL, Alfred George (1884-1968)

Gabriel worked for many years in the Department of Entomology of The Natural History Museum, London as Senior Scientific Officer. He published many entomological papers between 1936 and 1954, including: *Catalogue of the Type Specimens of Lepidoptera Rhopalocera in the British Museum (Natural History)*, co-authored with N.D. Riley (1924-1927), and the Rhopalocera part (Vol. 3, No. 3, 1939) *of Ruwenzori Expedition 1934-35*. He was a Fellow of the Royal Entomological Society of London.

Biographical Reference: 1969. Kennedy, J.S. *Proceedings of the Royal Entomological Society of London* (C) **33**:55.

Collection

(1) One Ms notebook: *Collections dealt with by A.G.G.*, c. 1923-1938.

146. GAEDE, Max

Primarily a Lepidopterist, he contributed parts to the work *Lepidopterorum Catalogus* (1911-1939), edited by P.O.C. Aurivillius and H. Wagner.

Collection

(1) One loose leaf Ms: [Unpublished Microlepidoptera parts of *Lepidopterorum Catalogus*], c. 1930s.

147. GAMBLES, Robert Moylan (1910-1990)

Born in London, Gambles was educated at Westminster School, Trinity College, Cambridge and the Royal Veterinary College, Edinburgh. Apart from a two-year spell in Palestine, he was Veterinary Officer in Cyprus, 1935-1948, and then served in Nigeria, 1949-1962, becoming Principal of the Veterinary School at Vom, finally returning to England in 1962. The last eight years of his career were spent as Research Officer with the Ministry of Agriculture, Fisheries and Food. His main interests were the Odonata, and the helminthological and entomological aspects of parasitology. Gambles was a Fellow of the Royal Entomological Society of London and an Honorary Associate of The Natural History Museum, London.

Biographical Reference: 1991. Parr, M.J. *Odonatologica* **20**:369-373.

Collection

(1) Four boxes: correspondence between Robert Gambles and other scientists, including:

Corbet, Phillip S.	1953-1986
Corbet, Sarah A.	1970-1984
Fraser, Frederick Charles (1880-1963)	1959-1962
Gardner, A.E.	1954-1976
Leonard, J.	1977-1985
Longfield, Cynthia (1896-1991)	1950-1985
Marshall, A.G.	1974-1980
Pinhey, Elliot Charles Gordon (b. 1910)	1959-1984

(2) Two boxes: Miscellaneous correspondence, including letters relating to Nature Conservancy Council Surveys, c. 1967-1988.

(3) One box: [Notes and diagrams on a variety of Odonata species, and helminth parasites].

(4) One box: Five Ms files, including, *Cyprus Laboratory Reports 1935-1938, Laboratory Reports 1939-1944* and *Colonial Records* which includes letters from R.M. Gambles in Nigeria.

(5) One box: Personal correspondence, 1965-1988.

(6) One box: [Miscellaneous loose notes, sketches and photographs].

(7) Nine boxes: Approximately 40 Ms drafts of scientific papers, c. 1960-1980.

(8) Typed Ms: [Lists of papers published by R.M. Gambles and a resumé of his education, career and qualifications], c. 1980.

148. GARDINER, J.

Collection

(1) Three Ms notebooks: notes on bee-keeping, and the sightings of various British insects, c. 1901-1963.

149. GEBIEN, Hans (1874-1947)

Born in Germany, he became a world authority on the Tenebrionidae. His collection was given to the Hamburg Museum and the Frey Collection in Munich.

Biographical Reference: 1948. Anon. *Coleopterists' Bulletin* **2**:25.

Collection

(1) Annotated typed Ms : *Catalogue of the Tenebrionidae of the Ethiopian Region* (Africa), 1947.

150. GENÉ, Carlo Giuseppe (1800-1847)

Educated at the University of Pavia, he studied philosophy and mathematics. He suffered a long and serious illness and during his convalescence became interested in natural history. Gené corresponded with Franco Andrea Bonelli (1784-1830) and when Bonelli died, was invited to take up his post in Turin. In 1821 he started to publish works on economic entomology. His later works concentrated on the taxonomy of various groups of insects and ticks.

Biographical Reference: 1851. Sismonda, E. *Memorie della reale Accademia delle Scienze di Torino* **11**:1-19.

Collection

(1) One Ms notebook: *Miscellanee entomologiche raccolte per cura ed uso del Dr G. Gené* [Extracts from papers by various entomologists, including Amédée Louis Michel Lepeletier de St Fargeau (1770-1845) and Dominique Cirillo (1734-1799)], c. 1830.

151. GIBBINS, Ernest Gerald (1900-1942)

In 1929 Gibbins joined the Uganda Malaria Unit and for the rest of his life worked on insects of medical importance, especially tsetse flies and the Simuliidae. He was murdered in Uganda by local tribesmen suspicious of his collecting activities. He was a Fellow of the Royal Entomological Society of London.

Biographical Reference: 1944. Cockayne, E.A. *Proceedings of the Royal Entomological Society of London* (C) **8**:70.

Collection

(1) Correspondence collection: 100 letters sent to Gibbins, 1936-1942, including letters from:

Barnard, Keppel H.	1937-1938
De Meillon, B.	1936-1939
Edwards, Frederick Wallace (1888-1940)	1932-1937
Marshall, Guy Anstruther Knox (1871-1959)	1940-1941

(2) Two Ms notebooks: [Notes on the Simuliidae], c. 1920-1942.

152. GODMAN, Frederick DuCane (1834-1919)

Born at Park Hatch, Surrey, Godman was educated at Eton and Trinity College, Cambridge. At university he met Osbert Salvin (1835-1898), who became his lifelong friend. They shared a love of natural history, especially ornithology, and were two of the founder members of the British Ornithological Union. In 1857 Salvin made the first of many visits to Central America and became fascinated by the natural history of the area. In 1861 Godman joined Salvin on his third expedition to Guatemala, collecting and observing natural history. Godman also became interested in island faunas and visited the Azores in 1865, publishing *The Natural History of the Azores* in 1870. When in London, Godman and Salvin spent much time together, studying their extensive collections and attending scientific meetings. They employed expert collectors, such as George Charles Champion (1851-1927), to visit special localities in many countries, including Colombia, Ecuador and Peru. In 1876 it was suggested that they should publish a fauna and flora of Central America and thus became joint editors of the classic work, *Biologia Centrali-Americana*, (1879-1915), consisting of 215 parts in 63 volumes. Many authors contributed to the work. The volumes concerned with the Aves and Rhopalocera were written by the editors. Salvin died in 1898, leaving Godman to complete the work. Godman purchased many collections of birds and insects, including the H.W. Bates collection of butterflies, the A. Sallé collection of Coleoptera and the Janson collection of Elateridae. All the insects from Mexico and Central America and the bird collections were presented to The Natural History Museum, London between 1870 and 1919. The butterfly collection alone consisted of over 100,000 specimens. The collections represent one of the most valuable acquisitions ever received by the Museum.

Godman was a Fellow of the Royal, Linnean and Zoological Societies. He was President of the Entomological Society of London and was a Trustee of The Natural History Museum, London. On his death, the Godman Trustees set up a fund to aid travel and research associated with natural history.

Biographical References: 1915. Godman, F.O. *Biologia Centrali-Americana* **1**:1-12, Autobiography; 1933. Neave, A. *Centennial History of the Entomological Society of London*.xlvi + 224pp. (pp.153-154.) London.

Collection

(1) One Ms notebook : *Mexican Rhopalocera in the Berlin Museum* (includes the Staudinger collection).
(2) One Ms notebook : *Index to the species and genera of American Hesperiinae in the Godman and Salvin Collection (some synonyms not included). The Pamphilinae have not been indexed*, c. 1901.
(3) Pen-and-ink and watercolour drawings from volumes of the *Biologia Centrali-Americana*:
- (i) *Diptera*, Vols. I-III (1886-1903): approximately 200 drawings by Frederick Maurits Van der Wulp (1818-1899) and J.E. Collin.
- (ii) *Rhopalocera*, Vol. III (1876-1901): 75 drawings of genitalia.
- (iii) *Hymenoptera*, Vol. III (1899-1901): approximately 50 drawings.
- (iv) *Orthoptera*, Vols. I-II (1893-1909): approximately 50 drawings.

(v) *Neuroptera*, (1892-1980): approximately 30 drawings.
(vi) *Coleoptera*, Vol. II (1887-1890) and Vol. IV (1884-1910): approximately 500 drawings by Baron Max Schlereth.
(vii) *Rhynchota*, Vol. II (1894-1909): approximately 300 drawings.

(4) One notebook: [lists stocks of copies of *Biologia Centrali-Americana*, 1907].
(5) One Ms leaf: *Number of insects presented by Dr. Godman up to Oct. 7th 1907.*

Insect Room Lists: Volume 1, item 35.

153. GOODSON, F. W.

Goodson was employed by Lionel Walter Rothschild (1868-1937) to work on his Lepidoptera collection at Tring Museum, Hertfordshire.

Collection

(1) One loose-leaf Ms: [notes on Lepidoptera], c. 1920s.
(2) One Ms notebook: *Notes on Theclinae, Fabricius*, c. 1920s.

154. GORHAM, Henry Stephen (1839-1920)

Educated at Rugby, he then studied civil engineering. He later studied theology at Lichfield and held a number of curacies before becoming vicar of Shipley, Sussex. He collected birds eggs, fossils, minerals and land-shells but his particular interest was Coleoptera. He published entomological papers for over 40 years, his most important works included *Endomycici recitati* (1873) and his contributions on the Erotylidae, Endomychidae and Coccinellidae (1887-1889) in the *Biologia Centrali-Americana* (1879-1915). Part of his Coleoptera collection was purchased by The Natural History Museum, London. B.M.1927-143. Gorham was a Fellow of the Royal Entomological Society of London, the Zoological Society and of the Société Entomologique de France.

Biographical Reference: 1920. Tomlin, J.R. le B. *The Entomologist's Monthly Magazine* **56**:112-113.

Collection

(1) Three volumes: published papers, c. 1840-1910, on Cleridae, Lampyridae and Lycidae, annotated by Gorham.
(2) Author's annotated copy: *Endomycici recitati: A Catalogue of the Coleopterous group, Endomycici, with Descriptions of New Species and Notes* (1873).
(3) Portrait in oils of H.S. Gorham by A.H. Smith, c. 1910.
(4) Portrait in oils of Clara d'Orville Gorham (1833-1915) by A.H. Smith, c. 1910.

155. GOSSE, Philip Henry (1810-1888)

Born at Worcester, Gosse developed a wide interest in natural history. He is remembered for his many popular works, including, *The Birds of Jamaica* (1847); *A Naturalist's Sojourn in Jamaica* (1851); *A Manual of Marine Zoology for the British Isles* (1855-56); *A Year at the Shore* (1865) and *The Great Atlas Moth of Asia* (1879). He published many entomological papers in *The Entomologist* and *The Zoologist*. In 1856 he was elected a Fellow of the Royal Society and in 1879 a Fellow of the Entomological Society of London.

Biographical References: 1888. Carrington, J.T. *The Entomologist* **21**:264; 1980. Freeman, R.B. and Wertheimer, D. *Philip Henry Gosse : A Bibliography,* 148pp. Dawson, Kent.

Collection

(1) One Ms notebook : *Descriptions of the Genital Prehensors in the Papilionidae, 1882.*

156. GOULD, Maurice H.

Collection

(1) Unpublished Ts: *Insectiana,* c. 1941. Includes five original oil paintings.

157. GRAVES, Philip Percival (1876-1953)

Educated at Oxford University, Graves became the Near East correspondent for *The Times*, London. During World War I he served in the Intelligence Corps in Egypt, Arabia and Palestine. After the war he returned to *The Times* and travelled widely in Ireland, India and the Balkans. He published many works on European and Near East problems. He developed a great interest in entomology, especially in the Lepidoptera and Odonata, collecting in Palestine, Cyprus, Bulgaria, Egypt, England and Ireland. He presented specimens to The Natural History Museum, London and in 1937 his collection of Lepidoptera and notebooks from the Near East were purchased by The Natural History Museum, London. B.M.1937-655. Graves was a Fellow of the Royal Entomological Society of London.

Biographical Reference: 1953. E.S.A.B. *The Entomologist's Record and Journal of Variation* **65**:272.

Collection

(1) Four Ms notebooks: [Inventory of 9000 Lepidoptera specimens from the Near East], c. 1937.
(2) One Ms notebook: *List of the Lepidoptera of the Near East in the B.M. Coll.*

158. GREEN, Edward Ernest (1861-1949)

Born in Ceylon, he was educated at Charterhouse School. He returned to Ceylon in 1880 to manage his father's plantations. Green developed a keen interest in mycology and entomology related to coffee and tea growing. On his retirement in 1913 he returned to England and expanded his collections of scale-insects. His publications included *The Coccidae of Ceylon* (1896-1922) and over 200 papers.

Biographical Reference: 1949 Laing, F. *The Entomologist's Monthly Magazine* **85**:215-216, portrait.

Collection

(1) One letter: from Green to G.F. Hampson regarding an insect drawing 1905.

Insect Room Lists: Volume 1, item 15.

(2) Original drawings for *The Coccidae of Ceylon* (1896-1922).

159. GUENÉE, Achille (1809-1880)

Born at Chartres, Guenée developed an early interest in entomology. His particular interest concerned the Lepidoptera and he became one of the most distinguished Lepidopterists of France. He published over 50 papers, mostly in the *Annales de la Société Entomologique de France.* His most important works were the 5 volumes (1852-1857) of the *Spécies Général des Lépidoptères* which formed part of *Collection des Suites à Buffon*.

Biographical References: 1881. Anon. *The Entomologist's Monthly Magazine.* **17**:214-216; 1881. Grote, A.R. *Papilio* **1**:31-35.

Collection

(1) One Ms notebook: *Catalogue général des Lépidoptères du globe*, c. 1860.
(2) One Ms notebook: [Observations on the synonymy of European Lepidoptera with an additional section giving provisional names of exotic species], c. 1857.
(3) One loose-leaf Ms: *Catalogue des Lépidoptères de ma collection. Classés systématiquement d'après ma méthode avec l'indication des genres et groupes qui me manquent de manière a présenter un Catalogue général des Lépidoptères du Globe*, 1863.

160. GUÉRIN-MÉNÉVILLE, Félix Édouard (1799-1874)

Born at Toulon, he developed an interest in all aspects of entomology. His early studies concentrated on descriptive and systematic work. Later he devoted much of his time to study of the the vine-pest *Phylloxera* and silk-producing moths. He founded the *Magasin de Zoologie* in 1831 and was editor of the journal until his death. He was an honorary Fellow of the Entomological Society of London.

Biographical Reference: 1874. Anon. *The Entomologist's Monthly Magazine* **10**:233-234.

Collection

(1) One Ms leaf: *Collection Guérin Ménéville Catalogues des types* .

Insect Room Lists: Volume 2, item 104.

(2) Original drawings and Ms text for an unpublished work on Agricultural Entomology.

161. GUNN, Donald Livingstone (1905-1983)

Donald Gunn graduated from Cardiff University in 1935. He moved to Birmingham University and remained there until 1946, carrying out pioneering work on developing an analytical approach to animal behaviour. In 1940 he co-authored with G.S. Fraenkel, *The Orientation of Animals*. In 1946 he joined the Anti-Locust Research Centre and began research on locust physiology and helped to encourage the use of aircraft in locust control. In 1952 Gunn became Director of the International Red Locust Control Service based in Northern Rhodesia and later Director of the Tea Research Institute in Sri Lanka. He was a Fellow of the Royal Entomological Society of London and was President of the Association of Applied Biologists.

Biographical Reference: 1984. Chapman, R.F. and Haskell, P.T. *Antenna* **8**:14-15.

Collection

(1) Three annotated photograph albums: [Photographs of research into locust control in Kenya], 1945.
(2) Ts: *Locust Encompassed*, c. 1950.
(3) Collection of miscellaneous Gunn reports, photographs and letters.

162. HALFORD, Frederick M. (d. 1914)

The Halford family was from the Midlands, but moved to London when Frederick was seven years old. Known for his important work on dry-fly fishing, Halford was also a keen entomologist. The vast amount of information which he amassed on aquatic insects was incorporated in his book *Dry Fly Entomology* (1902). Halford's other works include, *Floating Flies and How to Dress Them* (1886), *Dry Fly Fishing in Theory and Practice* (1889), *The Dry-Fly Man's Handbook* (1913) and *Modern Development of the Dry Fly* (1911). He also published articles in *The Field* under the pseudonym 'Detached Badger'.

Biographical Reference: 1903. Halford. F.M. *An Angler's Autobiography.* 286pp. London.

Collection

(1) One Ms volume: An incomplete draft of *The Dry-Fly Man's Handbook* published in 1913. Inscribed in the front, "Presented to Martin E. Mosely by Mr. Frederick M. Halford 1913."

163. HAMPSON, George Francis Bart. (1860-1936)

Hampson developed his entomological interests while engaged in tea planting in the Nilgiri Hills, South India. He started work in The Natural History Museum, London as a voluntary worker in 1889. He was the author of four volumes on moths (1892-1896) in the series *Fauna of British India* and later worked on the *Catalogue of the Lepidoptera Phalaenae in the British Museum,* published in 13 volumes (1898-1913), with 2 supplements (1914, 1920).

Hampson was a Fellow of the Royal Entomological Society of London.

Biographical Reference: 1937. Imms, A.D. *Proceedings of the Royal Entomological Society of London.*(C) **1**:55-56.

Collection

(1) Three Ms notebooks: *Nilgiri Diary,* 1888.

(2) Loose-leaf Ms: *A Classification of the Pyralidae subfamily Anerastianae,* c. 1930s.

(3) Correspondence collection: approximately 20 letters sent to and from Hampson, 1905-1930.

(4) Three boxes: approximately 1000 original drawings for the Moth volumes of *Fauna of British India* (1892-1896) and the *Catalogue of the Lepidoptera Phalaenae in the British Museum* (1898-1913).

(5) One Ms notebook: *List of Phycitinae as left in collection by Hampson*, by R. J. West.

(6) Miscellaneous notes and reprints.

(7) Ms notes relating to *Catalogue of the Lepidoptera Phalaenae in the British Museum and supplements.*

(8) Two Ms notebooks : *Notes on types of genera*. Contains references to original descriptions of genera.

(9) Eleven loose-leaf Ms files: [Catalogue of Noctuidae Ophiderinae], c. 1920.

(10) One loose-leaf Ms: [Key to the genera of Noctuidae], c. 1920.

(11) One loose-leaf Ms: [Phylogenetic table and key].

(12) Five loose-leaf Ms files: [Catalogue of the Noctuidae Hypeninae], c. 1920.

164. HANCOCK, E. Geoffrey (b. 1948)

Formerly with the Bolton Museum, Hancock is now Keeper of Natural History at the Kelvin Grove Museum, Glasgow. A keen Dipterist and all-round naturalist, he was co-editor of *A Survey of Zoological and Botanical Material in Museums and Other Institutions in Great Britain* (1980).

Collection

(1) One Ts: *Notes on Rye Types in Bolton Museum*. Notes on the type specimens of Edward Caldwell Rye (1832-1885), published in 1985 under the title *Rye's Beetles: A Catalogue of E.C. Rye's Type Specimens in Bolton Museum.*

165. HANCOCK, George Leonard Rhys (b. 1940)

Educated at Harrow School and Cambridge University, he developed a keen interest in entomology. In 1926 he was appointed Assistant Agricultural Entomologist to the Uganda Government, working especially on pests of cotton. He was also interested in mosquitoes and worked with G.H.E. Hopkins.

Later Hancock became lecturer at Makerere College. He was a member of the East Africa and Uganda Natural History Society and a Fellow of the Royal Entomological Society and the Zoological Society of London.

Biographical Reference: 1940. Gedge, A.F.J. *Journal of the East Africa and Uganda Natural History Society*. **15**:83-84.

Collection

(1) One loose-leaf Ts: [statistics for The mosquitoes of Namanve swamp, Uganda, *Journal of Animal Ecology* **3**:204-221 (1934).]

166. HAPPE, Andreas Friedrich (1733-1802)

Happe was an accomplished German naturalist. His interests covered a range of natural history subjects, including entomology and botany. Some of his published works have become extremely rare, including the entomological work *Abbildungen der Schmetterlinge* (1783-1785).

Biographical References: 1969. Sawyer, F.C. *Journal of the Society for the Bibliography of Natural History* **5**:123-124; 1953. Rautenberg, L.E. *Beiträge zur Entomologie* **13**:618-626.

Collection

(1) Six bound volumes: 840 original watercolour drawings of insects, arachnids and crustaceans with Ms notes *Naturgeschichte der Insekten* (1773). This work was unpublished.
(2) One bound volume: original watercolour drawings, mostly exotic Lepidoptera, c. 1750.

167. HAUSER, Friedrich (1853-1932)

Born in Nördlingen, Germany, he developed a special interest in Palaearctic Coleoptera and Lepidoptera. In 1904 his extensive collection of Central Asian Coleoptera was purchased for The Natural History Museum, London. B.M.1904-63.

Biographical Reference: 1935. Ross, E. *Intenationale Entomologische Zeitschrift* **28**:484.

Collection

(1) Six Ms leaves: [Lists of the 4,000 Coleoptera specimens in the Hauser Collection. B.M.1904-63].

Insect Room Lists: Volume 2, item 82.

168. HAVILAND, George D. (1857-1901)

Haviland developed a special interest in termites. In 1898 he published observations on these insects with descriptions of new species (*Journal of the Linnean Society* **26**:358-442). In 1899 5,000 specimens from Haviland's Collection came to The Natural History Museum, London. B.M.1899-41.

Collection

(1) One Ms leaf: *Haviland Collection of Termites for the British Museum*, 1899.

Insect Room Lists: Volume 2, item 105.

169. HEALY, C.

Collection

(1) One Ms notebook: *General observations on the genus* Gracilaria, c. 1850.

170. HEDGES, Alfred Vander (1893-1957)

Hedges developed an interest in entomology from an early age. His particular area of study was breeding Lepidoptera. He was able to inbreed several species such as *Lygris testata* L. and *L. populata* L. and discovered rare recessives in the populations. His collection of 20,000 British moths was bequeathed to The Natural History Museum, London in 1958. B.M.1958-441. Hedges was a Fellow of the Royal Entomological Society of London.

Biographical Reference: 1958. H.B.D.K. *The Entomologist* **91**:51, portrait.

Collection

(1) Five collecting notebooks: 1920-1957.
(2) Correspondence collection: concerns the Hedges bequest of specimens, 1951-1961.

171. HEMMING, Arthur Francis (1893-1964)

Educated at Rugby, Hemming read history at Corpus Christi College, Oxford. He was wounded during World War I and in 1918 joined the Civil Service. He served as private secretary to various ministers and was awarded the CMG and the CBE for his services. Hemming also served as Secretary to the International Commission for Zoological Nomenclature, launching the *Bulletin of Zoological Nomenclature* and helping to improve the financial structure of the Commission. His special entomological interest concerned Lepidoptera, on which he published many papers. In 1934 he published *Generic Names of the Holarctic Butterflies, 1758-1863* and, in 1937, *Hübner: A Bibliographical and Systematic Account of the Entomological Works of Jacob Hübner and of the supplements thereto by Carl Geyer, Gottfried Franz von Frölich and Gottlieb August Wilhelm Herrich-Schäffer.* During his lifetime Hemming published over 1,000 titles.

Biographical References: 1964. Riley, N.D. *Bulletin of Zoological Nomenclature* **21**:402-404, portrait; 1964. Riley, N.D. *Journal of the Lepidopterist's Society* **18**:237-239.

Collection

(1) One box loose-leaf Ts: *Generic Names of the Butterflies of the World*, c. 1938-1964.
(2) Three loose-leaf Ms notebooks: [Information on Lepidoptera species].
(3) One folder of Ts:
    (i) *Summer holidays, 1959. Itinerary and captures in and around Ille-et-Vilaire, France.*
    (ii) *Catalogue of the Entomological Library formed by the Late A.F. Hemming*, 1964.

(4) Fifteen loose-leaf Ms notebooks: [references to Lepidoptera].

(5) Correspondence collection: approximately 50 letters relating to the International Commission on Zoological Nomenclature, mostly between Prof. C.W. Stiles and F.A. Bather, c. 1922.

(6) One loose-leaf Ts: *Alphabetical index of the species, the types of which are figured in Jones' Icones.*

(7) Three miscellaneous entomological notebooks.

(8) One Ms notebook: *Hesperiidae.*

(9) One box of loose-leaf Ts:

- (i) [Working papers on Lepidoptera names].
- (ii) [H] *Stempffer's paper on African Lycaenidae.*

(10) Five boxes of Mss:

Box 1: Loose-leaf Ms relating to generic names of butterflies.

Box 2: Loose-leaf Tss:

- (i) *Linné's entomological lecture notes, 1755.*
- *(ii) Systema Papilionum palaearcticorum.*
- (iii) *Rhopalocera - Family Group problems.*
- (iv) *Names for North American Hesperiids.*
- (v) *Generic names of North American Butterflies for the Official List.*

Box 3 : Loose-leaf Tss:

- (i) [References to original descriptions of Lepidoptera species].
- (ii) *Alphabetical Index of Generic names.*
- (iii) *A list, arranged alphabetically by families, of the species which are the types of Genera represented in the Holarctic Region, with references to the place in which each such species was first described or figured.*
- (iv) *A synonymic checklist of the genera of Holarctic butterflies with their type species and a list giving reference to the place in which each such species was first described or figured.*
- (v) *Revision of the dates of Vol.1 of the present work to generic and specific names published in the entomological works of Jacob Hübner and of the supplements thereto published by Carl Geyer.*

Box 4 : Loose-leaf Tss:

- (i) [Notes on Hesperiinae].
- (ii) *Genera of the Butterflies of the World.*
- (iii) *Notes on the publication dates of* [J.B] *Godart &* [P.A.J.] *Duponchel,*

Histoire Naturelle des Lépidoptères ou Papillons de France (1821-1840), c. 1930.

Box 5 : One box of loose-leaf Tss:

(i) *Generic Names of the Holarctic Butterflies: Revisional notes for Vol. II.*

(ii) [O.] *Bang-Haas, Horae macrolepidopterologicae, date of publication.*

(iii) *Checklist of European Butterflies.*

(iv) Correspondence with Roger Verity, c. 1948-1958.

(11) Two bound volumes of correspondence: approximately 200 letters, sent to and received from entomologists, 1919-1939, including:

Avinoff, André (1884-1949)	1935
Barnes, William (1860-1930)	1927
Bell, Ernest Layton (1876-1964)	1935
Benjamin, Forster Handrickson (1895-1936)	1933-1936
Bodenheimer, Frederick Simon (1897-1959)	1933
Burrows, Charles Richard Nelson (1851-1936)	1923-1926
Buxton, Patrick Alfred (1892-1955)	No date
Bytinski-Salz, Hanan (1903-1939)	1936
Corbet, Alexander Steven (1896-1948)	1934-1936
Edwards, Frederick Wallace (1888-1940)	1936
Evans, William Henry (1876-1956)	1928-1936
Fletcher, Thomas Bainbrigge (1878-1950)	1934
Forbes, William Trowbridge Merrifield (1886-1968)	1934
Fountaine, Margaret Elizabeth (1862-1940)	1936
Graves, Philip Percival (1876-1953)	1931
Holland, William Jacob (1848-1932)	1929-1930
Joannis, Joseph de (1864-1932)	1926-1927
Lindsey, Arthur Ward (1894-1963)	1934
McDunnough, James Halliday (1877-1962)	1931
Peile, Harry Diamont (1872-1959)	1927-1936
Poulton, Edward Bagnall (1856-1943)	1926-1935
Querci, Orazio (1875-1967)	1922-1934
Rebel, Hans (1861-1940)	1928
Seitz, Adalbert (1860-1938)	1928
Stiles, Charles Wardell (1867-1941)	1935

Turati, Emillo (1858-1938)	1928
Turner, Henry Jerome (1856-1951)	1926-1928
Verity, Roger (1883-1959)	1925-1934
Wagner, Fritz (1873-1938)	1936
Waterhouse, Gustavus Athol (1877-1950)	1934-1936
Zerny, Hans (1887-1945)	1928-1932

(12) Correspondence collection: approximately 600 letters sent to and received from entomologists, c. 1942-1959, including:

Clark, Austin Hobart (1880-1954)	1950
Ford, Edmund Bristo (1902-1988)	1957
Higgins, Lionel G. (1891-1985)	1950-1957
Neave, Sheffield Airey (1879-1961)	1940
Querci, Orazio (1875-1967)	1958
Toxopeus, Lambertus Johannes (1894-1951)	1948
Warren, Brisbane Charles Somerville (1887-1979)	1958

(13) One box miscellaneous notes and letters.
(14) Six Ms volumes: *The Butterflies of France. Distribution by Departments*, c. 1950.

172. HERING, Erich Martin (1893-1967)

Born at Heinersdorf, Germany, he was originally a primary schoolteacher. Later he attended university, gained his Ph.D. and joined the Zoologisches Museum der Universität Berlin. His particular interest was leaf-mining insects, and his major work, *Biology of the Leaf-Miners*, was published in 1951. He was a member of the International Commission for Zoological Nomenclature, an honorary member of British, Belgian and French entomological societies, The Lepidopterists' Society, and the International Entomology Club in Frankfurt. He won the Fabricius Medal in Germany, and received the Royal Bulgarian Medal of Merit and the Royal Belgian Medal for Natural Sciences.

Biographical Reference: 1968. Hannemann, H.J. *Journal of the Lepidopterists' Society* **22**:123-124.

Collection

(1) One Ms notebook: *Katalog der Minen-Sammlung - Band I: Minen aus dem Palaearctischen.*
(2) Three Ms notebooks: [Palaearctic leafminers], c. 1960.

173. HERON, Francis Arthur (1864-1940)

Heron was educated at Charterhouse School and New College, Oxford. From 1889 until his retirement for health reasons in 1910, he worked at The Natural History Museum, London in the Entomology Department. From 1901 onwards he was Assistant-in-Charge of Butterflies. In 1909 he published 'Lepidoptera Rhopalocera, Ruwenzori Reports', *Transactions of the Zoological Society of London* **19**:141-178. He was a Fellow of the Entomological Society of London.

Biographical Reference: 1942. A.G. G[abriel]. *The Entomologist* **75**:23-24.

Collection

(1) One Ms notebook: notes, and reprints from journals, c. 1890-1903.

174. HEWER, Humphrey Robert (1903-1974)

Educated at Imperial College, London, Hewer joined the staff of the College in 1926, and became Professor of Zoology in 1964. In World War II he was seconded to the Ministry of Agriculture, Fisheries and Food, and in 1965 he became Chairman of the MAFF Infestation Control Laboratory Advisory Committee. He was Vice-President of the Council for Nature, and a member of the Committee for England of the Nature Conservancy. He was awarded a CBE in 1971.

Biographical Reference: 1974. Anon. *Habitat* **10**(3):3.

Collection

(1) Loose-leaf Ms notes: *Genitalia Data* Zygaena.
(2) Loose-leaf Ms notes: *Statistical Data* Zygaena.
(3) Loose-leaf Ms notes: *Genitalia Data-Foreign* Zygaena.
(4) Loose-leaf Ms notes: *Genitalia Data-British* Zygaena.
(5) Loose-leaf Ms notes: *Wing Drawings*. Notes and drawings on the wings of *Zygaena*.
(6) One Ms notebook: *Serial Catalogue*. Ms catalogue of *Zygaena* collection, arranged in order of serial numbers.
(7) Two Ms notebooks: [Details of British and European *Zygaena* colonies.]
(8) One Ms notebook: *Collecting Diary kept at Digne, 1929.*
(9) One Ms notebook: *Old Testament Criticism Continued - Reconstruction Contd*. Includes entomological notes in the second half of the volume. c. 1928-1929.
(10) One Ms notebook: *Bethune-Baker Collection.*
(11) Miscellaneous Mss: [Notes on *Zygaena*.]

175. HEWITSON, William Chapman (1806-1878)

Born at Newcastle-on-Tyne, he practised as a surveyor in York and Bristol. Hewitson was especially interested in Lepidoptera, and published numerous papers. He was a very skilled artist and is particularly well known for his plates in *The Genera of Diurnal Lepidoptera* (1846-1852), by E. Doubleday and J.O.Westwood.

Biographical Reference: 1879. Anon. *The Entomologist's Monthly Magazine* **15**:44-45.

Collection

(1) Correspondence collection: photocopies of approximately 60 letters c. 1870-1878 from Hewitson to Herman Strecker (1836-1902). The originals are held in the Colorado State Archives.
(2) Photocopy of the last will and testament of William Chapman Hewitson, 1878.

176. HICKIN, Norman Ernest (1910 -1990)

Born in Birmingham, he was educated at the University of London and spent most of his professional life with Rentokil, a pest control and wood preservation company. His specialism was wood-boring insects, especially beetles and termites. In 1985 he published *Longhorn Beetles of the British Isles.* He was the Scientific Director of Rentokil Limited for 30 years, and a founder member of the Society of Wildlife Artists.

Biographical Reference: 1991. Bateman, P.L.G. *Antenna* **15**:60-61.

Collection

(1) Nine Ms diaries and notebooks, 1937-1957.

177. HIGGINS, Lionel George (1891-1985)

Born in Bedford, he was educated at Cambridge and qualified as a doctor, later specialising as an obstretician and gynaecologist. His entomological interests were the taxonomy and systematics of Palaearctic butterflies. Higgins bequeathed his large collection, particularly rich in Palaearctic Nymphalidae, to The Natural History Museum, London. B.M.1985-476. He published many important works, including *The Classification of European Butterflies* (1981) and, with N.D.Riley, *A Field Guide to the Butterflies of Britain and Europe* (1970). He was an Honorary Associate of The Natural History Museum and held Honorary Membership of the British Entomological and Natural History Society, The Lepidopterists' Society and Societas Europaea Lepidopterologica. He served on the Council of the Royal Entomological Society of London.

Biographical Reference: 1986. Gilbert, P., Tremewan, W.G. and Vane-Wright, R.I. *Antenna.* **10**:65-67, portrait; 1988. Smith, C.R. *Entomologist's Gazette* **39**:17-37, bibliography.

Collection

(1) One box of correspondence and notes, 1978-1985, concerning the

depositing of the original drawings for *A Field Guide to the Butterflies of Britain and Europe* (1970) in The Natural History Museum, London, and correspondence on the sale of part of the Higgins Library to The Natural History Museum, London.

(2) Three boxes of Ms notes concerning *A Field Guide to the Butterflies of Britain and Europe* (1970) by L.G. Higgins and N.D. Riley, with illustrations by Brian Hargreaves.

(3) One box of Ts and Ms notes: concerns the second edition of *A Field Guide to the Butterflies of Britain and Europe* (1973), by L.G. Higgins and N. D. Riley, with illustrations by Brian Hargreaves.

(4) Three Ms notebooks: [Lists of numbered preparations of butterfly genitalia, c. 1915-1982.]

178. HILLE RIS LAMBERS, Dick (1909-1984)

Educated at the Secondary Colonial Agriculture School, Deventer, The Netherlands, he also studied at the University of Wageningen but did not take a formal degree. He received an honorary doctorate from the University of Milan in 1967. His main interest was aphids, he was joint author with V.F. Eastop of, *Survey of the World's Aphids* (1976). He was an Honorary Fellow of the Royal Entomological Society, and was elected a Foreign Member of the Linnean Society in 1981.

Biographical Reference: 1984. Stroyan, H.L. *Antenna*. **8**:127-129, portrait.

Collection

(1) Six field notebooks, 1939-1984.

179. HINTON, Howard Everett (1912-1977)

Born in Mexico, Hinton was educated at the University of California and at King's College, Cambridge. He took part in scientific expeditions to Mexico, 1933, and to Peru, Bolivia and Brazil, 1937. He worked at The Natural History Museum, London from 1939 to 1949, when he was appointed Lecturer in the Department of Zoology at Bristol University. He was later promoted to

Reader and in 1970 became Head of the Department of Zoology. His particular interest was Coleoptera, and he made extensive use of the electron microsope to study eggs, spiracles and cuticular structures in beetles and other insects. His three-volume work, *Biology of Insect Eggs* was published posthumously in 1981. He founded and edited the periodical *Journal of Insect Physiology*. Hinton was a Fellow of the Royal Society, and President of the Royal Entomological Society, London.

Biographical Reference: 1978. Sait, G. *Biographical Memoirs of Fellows of the Royal Society* **24**:151-182, bibilography, portrait,

Collection

(1) One Ts: *Biology of the Lepidoptera* (unpublished).

180. HOCKING, John H.

A lepidopterist who published various notes in periodicals, including 'Notes on Indian Butterflies' in *Science Gossip* **18**: 271 (1882).

Collection

(1) One Ms notebook and two Ms lists: [Hocking's Lepidoptera Collection].
(2) One Ms notebook: *Memoranda of larvae etc.* Concerning Hocking's collection of larvae from Dharmsala, India, c. 1880.

181. HOFMANN, Ottmar (1835-1900)

Born in Frankfurt, he moved with his parents to Regensburg in 1846. He became a physician in the Bavarian State Service. Hofmann's main entomological interest was Lepidoptera and he published many papers on the Tineids. His collections came to The Natural History Museum, London via Lord Walsingham (1843-1919) B.M. 1910-427.

Biographical Reference: 1900. Walsingham. *The Entomologist's Monthly Magazine* **36**:212-213.

Collection

(1) One loose-leaf Ms: *Pflanzen und Raupen Microlepidoptera,* c. 1860-1870.
(2) One loose-leaf Ms: *Notizen . . . Psychiden,* c. 1858, notes for Ueber die Naturgeschichte der Psychiden, *Berliner Entomologische Zeitschrift* **4**:1-53 (1860).

182. HOLMES, J.W. O. (d. 1980)

A doctor of medicine and an enthusiastic amateur entomologist, he was particularly concerned with conservation. He was a member of the Hampshire and Isle of Wight Naturalists Trust, and a Fellow of the Royal Entomological Society.

Biographical Reference: 1981. Brewis, A. *Annual Report of the Hampshire and Isle of Wight Naturalists Trust Ltd* **1981**:5.

Collection

(1) One Ms notebook: miscellaneous notes, newspaper cuttings and correspondence relating to British entomology, c. 1936-1979.

183. HONRATH, Eduard G. (1837-1893)

Born in Coblenz, he worked as an art dealer in Berlin. Honrath was especially interested in Lepidoptera, and had a collection which was particularly rich in the genus *Parnassius*.

Biographical Reference: 1893. Elwes, H.J. *Proceedings of the Entomological Society of London* **1893**;lvii.

Collection

(1) One Ms catalogue: *Verzeichniss der Exotischen Lepidopteren der Collection Ed. G. Honrath.*

184. HOPKINS, George Henry Evans (1899-1973)

A senior medical entomologist at Kampala, Uganda, he was particularly interested in lice, fleas and mosquitoes. He published *Mosquitoes of the Ethiopian Region* (1936). Hopkins' large collection of Mallophaga was presented to The Natural History Museum, London, B.M. 1958-686 and B.M.1961-199.

Biographical Reference: 1974. Lees, A.D. *Proceedings of the Royal Entomological Society of London*. (C) **38**:60.

Collection

(1) One Ts: *Trichodectidae descriptions* (unpublished).

185. HOWARTH, Thomas Graham

Howarth specialized in Lepidoptera, and worked in the Department of Entomology of The Natural History Museum, London from 1936 to 1976, although his career was temporarily interrupted by World War II. During the War, he spent time in a Japanese prisoner of war camp in Malaysia, but continued to collect Lepidoptera.

His publications include *South's British Butterflies : Based Extensively on the Classic by Richard South* (1973).

Collection

(1) One Ts: *A list of the Frühstorfer collection*. A 34-leaf catalogue of the Rhopalocera collections of Hans Frühstorfer (1863-1922), B.M.1933-131 and 1937-285.

186. HUDSON, C. W. M.

Collection

(1) One Ms notebook: *Notes on the Walsingham Collection of Moths*, 1939.

187. HUMPHRIES, E.C.

Collection

(1) One Ms notebook: lists 1029 Coleoptera from Trinidad, B.M.1953-636.

188. HUTCHINSON, Mrs.

Married to Rev. Thomas Hutchinson, she shared with him a love of natural history and the family were leading members of the Woolhope Naturalists' Field Club, Herefordshire. Their eldest child, Thomas Hutchinson (1848-1916) was a keen entomologist and ornithologist. In 1937, Miss Emma Hutchinson presented to The Natural History Museum, London a collection of over 20,000 Lepidoptera specimens, B.M.1937-472.

Biographical Reference: 1983. Miles, B.E. *Transactions of the Woolhope Naturalists' Field Club.* **43** (1981):336.

Collection

(1) One diary : *Mrs. Hutchinson, Diary 1.* Notes on the entomological collection of Thomas Hutchinson (1848-1916) for the years 1868-c.1893.
(2) One diary : *Diary 2.* Lists specimens presented to the Natural History Museum. B.M. 1937-472.

189. INGALL, Thomas (1800-1862)

Ingall was employed for many years by the Bank of England. He was a keen amateur naturalist, with a particular interest in insects. He was one of the original members of the Entomological Club and the Entomological Society of London.

Biographical Reference: 1895. South, R. *The Entomologist* **28**:40-45.

Collection

(1) One Ts: *The Irregular Diary of an Entomologist, 1832-1860*. A copy by his

son, Mr. W.T.F.M. Ingall. This formed the basis of the biographical article published in *The Entomologist.*

(2) Three letters from W.T.F.M. Ingall to Richard South (1846-1932) concerning the above entomological diary, 1894.

190. JACOBY, Martin (1842-1907)

Jacoby was a musician who played in the orchestra of the Royal Italian Opera; later he became a violin tutor. His interests turned to entomology and he developed a particular interest in the Coleoptera and especially phytophagous beetles. He was the author of the Phytophaga volume in F.D. Godman and O. Salvin's *Biologia Centrali-Americana,* Insecta, Volume 6 part 1 and Supplement part 1 (1888-1892). Jacoby was a Fellow of the Entomological Society of London.

Biographical Reference: 1908. Walker, J.J. *The Entomologist's Monthly Magazine* **44**:45.

Collection

(1) One Ms: *Alphabetical index to* [J.S] *Baly and Jacoby Collection.*

191. JAKUBSKI, Antoni Wladyslaw (1885-1962)

Born in Poland, he developed entomological and zoological interests. He served with the Polish Free Army during World War II and was based in Britain where he afterwards settled and worked in the Entomology Department of The Natural History Museum, London as an Honorary Associate. He worked on the Margarodidae and Termitococcidae, his research culminating in the published volume *A Critical Revision of the Families Margarodidae and Termitococcidae (Hemiptera : Coccoidea)* (1965).

Biographical Reference: 1988. Brzęk, G. *Przeglad Zoologiczny* **33**:137-158.

Collection

(1) Tss, drawings and photographs: all relate to *A Critical Revision of the Families Margarodidae and Termitococcidae (Hemiptera : Coccoidea)*. Also includes an unpublished bibliography of the above work, containing references dating back to Biblical times.

192. JANSON FAMILY ARCHIVE

In 1852 the Janson family established a business as Natural History agents and booksellers at 44 Great Russell Street, London. For over 90 years they were purchasers and sellers of natural history specimens and collections from around the world, corresponding with distinguished naturalists of the day. The archives of the Janson family were presented to the Natural History Museum in 1991 by D.B. Janson.

Collection

(1) The archive contains a vast amount of material including business papers, letters, letterbooks and account books, c. 1873-1929. It includes the papers of:

Edward Wesley Janson (1822-1891)
Oliver Erichson Janson (1850-1926)
Oliver Jordan Janson (1876-1964)
O.E. Janson & Son
Janson & Sons

The correspondence collection contains thousands of letters to and from entomologists and naturalists regarding the purchase and sale of collections, including specimens collected on expeditions and from private collections:

Adams, Herbert Jordan (1838-1912)	1898
Adkin, Benaiah Whitley (1865-1948)	1909-1921
Adkin, Robert (1849-1935)	1894-1927
Andrewes, Herbert Edward (1863-1950)	1899-1926
Angell, John Watson (1885-1946)	1910-1920
Armitage, Edward (1817-1896)	1872-1879

Arrow, Gilbert John (1873-1948)	1899-1922
Bagnall, Richard Siddoway (1889-1962)	1904-1908
Barrett, Charles Golding (1836-1904)	1899-1900
Bates, Frederick (1829-1903)	1872-1903
Bates, Henry Walter (1825-1892)	1872-79
Bell-Marley, H.W. (1872-1945)	1912-1917
Bethune–Baker, George Thomas (1857-1944)	1901-1928
Black, James Ebenezer (1865-1924)	1900-1924
Brölemann, Henri (1860-1933)	1913
Brunetti, Enrico Adelelmo (1862-1927)	1905-1913
Buckton, George Bowdler (1818-1905)	1898
Burr, Malcolm (1878-1954)	1898-1919
Butler, Edward Albert (1845-1925)	1908
Cameron, Malcolm (1873-1954)	1913-1927
Carpenter, Geoffrey Douglas Hale (1882-1953)	1928
Carrington, John Thomas (1846-1908)	1906
Carter, Herbert James (1858-1940)	1928-1929
Champion, George Charles (1851-1927)	1897-1926
Chitty, Arthur John (1859-1908)	1900
Christy, William Miller (1863-1939)	1898
Cockayne, Edward Alfred (1880-1956)	1911
Coe, Ralph Leonard (1906-1968)	1927
Collin, James Edward (1876-1968)	1899-1923
Cooke, Benjamin (1816-1883)	1874
Crawford, William Monod (1876-1942)	1923
Crotch, George Robert (1841-1874)	1872-1874
Davidson, James (1885-1945)	1922-1925
Day, George O. (1854-1942)	1896-1905
Distant, William Lucas (1845-1922)	1879-1920
Doherty, William (1857-1901)	1893-1901
Dow, Robert Percy (1865-1936)	1914
Dunning, Joseph William (1833-1897)	1872-1877
Elwes, Henry John (1846-1922)	1893-1894
Embry, B. (1898-1963)	1927
Fitch, Edward Arthur (1854-1912)	1874-1877

Fowler, William Weeks (1849-1923)	1877-1911
Froggatt, Walter Wilson (1858-1937)	1927
Frohawk, Frederick William (1861-1946)	1894-1922
Fryer, Herbert Fortescue (1854-1930)	1877-1879
Gimingham, Conrad Theodore (1884-1957)	1920-1926
Gorham, Henry Stephen (1839-1920)	1872-1911
Gosse, Philip Henry (1810-1888)	1879
Grandi, Guido (1886-1970)	1926
Gray, John Edward (1880-1875)	1872
Hallett, Howard Mountjoy (1878-1958)	1917
Hampson, George Francis (1860-1936)	1902-1906
Harold, Edgar von (1830-1886)	1871-1875
Harrison, Albert (1860-1911)	1904-1909
Heidemann, Otto (1842-1916)	1904
Hemming, Arthur Francis (1893-1964)	1919-1929
Hewitson, William Chapman (1806-1878)	1873-1876
Holland, William Jacob (1848-1932)	1893-1929
Horn, George Henry (1840-1897)	1873-1877
Jeffery, Hugh Geoffrey (d. 1943)	1923
Johnson, Charles Willison (1863-1932)	1904-1905
Joy, Norman Humbert (1874-1953)	1902-1920
Karny, Heinrich Hugo (1886-1939)	1929
Kirby, William Forsell (1844-1912)	1879-1905
Leng, Charles William (1859-1941)	1915
Lowther, Richard Charles (1884-1950)	1921-1924
Mason, Frank R. (1882 1927)	1909 1918
May, John William (1814-1902)	1875
Meyrick, Edward (1854-1938)	1877-1905
Navas, Longinos (1858-1938)	1923
Newman, Edward (1801-1876)	1872-1875
Oberthür, Charles (1845-1924)	1893-1924
Ormerod, Eleanor Anne (1828-1901)	1879-1900
Parry, Frederick John Sidney (1810-1885)	1873-1875
Pascoe, Frances Polkinghorne (1813-1893)	1875
Pic, Maurice (1890-1957)	1929

Pickard–Cambridge, Arthur (1873-1952)	1917
Reichensberber, August (1878-1962)	1909-1929
Reynolds, Lawrence R. (1878-1922)	1917-1918
Roebuck, William Denison (1851-1919)	1874-1907
Rye, Edward Caldwell (1832-1885)	1872-1875
Schmiedeknecht, Otto (1847-1936)	1928
Selous, Cuthbert Fennessy (d. 1946)	1906-1929
Sheldon, William George (1859-1943)	1896-1926
Shipp, John William (d. 1898)	1893-1897
Stainton, Henry Tibbats (1822-1892)	1872-1874
Stoneham, Hugh Frederick (1889-1966)	1929
Swinhoe, Charles (1836-1923)	1893
Theobold, Frederick Vincent (1868-1930)	1921
Wallace, Alfred Russel (1823-1913)	1876-1903
Wollaston, Thomas Vernon (1822-1878)	1872-1878

193. JEPSON, F. P. (d. 1950)

Jepson worked for a number of years as Government Entomologist in the Department of Agriculture, Fiji. He was then employed for 13 years in Ceylon as Assistant Entomologist. His work included studies on the Shothole borer of tea *Xyleborus fornicatus* and the tea termite *Calotermes*.

Biographical Reference: 1950. A.W.R. *Tropical Agriculturalist* **106**:130.

Collection

(1) One Ts: *List of Coleoptera collected in Fiji and sent to the British Museum for identification 1909*. B.M.1910-111.

Insect Room Lists: Volume 2, item 88.

194. JERDON, Thomas Claverhill (1811-1872)

Born in County Durham, Jerdon was educated at Edinburgh University. He subsequently qualified in medicine and served as a surgeon to the East India

Company. He became extremely interested in the natural history of India and published a *Catalogue of the Birds of the Peninsula of India* (1839).

Biographical Reference: 1876. Anon. *History of the Berwickshire Naturalist's Club,* **1873-75**:143-151.

Collection

(1) One Ms: *A Catalogue of the Species of Ants found in Southern India,* 1851. Published 1851 in *Madras Journal of Literature and Science* **16**:103.

195. JOICEY, James John (1871-1932)

From boyhood he was interested in Lepidoptera. In 1910 he acquired the Henley Grose-Smith collection, and three years later the Herbert Druce collection. Between 1913 and 1921 Joicey bought the following collections: Roland Trimen, 1916, Colonel Swinhoe, 1916, Lt.-Col. C. G. Nurse, 1919, Hamilton Druce, 1919 and Dognin 1921. The collections were kept at Joicey's home, The Hill Museum, Witley, Surrey and he added to it by sending special collectors to explore various regions on his behalf, e.g. the Pratt brothers to South America and New Guinea, and T. A. Barns to Central Africa. By 1930 the Hill Museum contained upwards of 380,000 specimens. He presented approximately 75,000 specimens to The Natural History Museum, London. As part of the Joicey Bequest, the Museum received a further 300,000 specimens, B.M.1934-120. Joicey published four volumes of *Bulletin of the Hill Museum*, 1921-1932. A *Catalogue of the Type Specimens of Lepidoptera Rhopalocera in the Hill Museum* was published by A.G. Gabriel in 1932.

Biographical Reference: 1932. Riley, N.D. *The Entomologist* **65**:142-144.

Collection

(1) Hill Museum Visitors' Book 1914-1932, including signatures of many well-known entomologists.
(2) Loose-leaf Ms: Lists of Lepidoptera specimens.

(3) Correspondence collection. Approximately 20 letters from:

Collenette, Cyril Leslie (1888-1959)	1927-1928
Oberthür, Charles (1845-1924)	1922
Poulton, Edward Bagnall (1856-1943)	1914

196. JONES, William (d. 1818)

A wealthy wine merchant, Jones painted from nature approximately 1500 species of butterflies and arranged them according to the Linnean classification. Many served as Fabrician types. These paintings, known as 'Jones Icones', are in the Hope Entomological Collections, Oxford. Because of the importance of the collection, attempts were made – unsuccessfully – to publish the paintings. In 1977 when the volumes were prepared for conservation, they were photographed and the photographs made available for purchase. A set of colour slides is now held in the Entomology Library of the Natural History Museum, London.

Biographical Reference: 1934. Poulton, E.B. *Transactions of the Society for British Entomology* **1**:139-149.

Collection

(1) One folder: [Correspondence and notes from G.D. Hale Carpenter (1882-1953), A.F. Hemming (1893-1964) and others concerning a proposed publication of the 'Jones Icones'.]

197. JORDAN, Heinrich Ernst Karl (1861-1959)

Born in Hannover, Jordan received his Ph.D. from Göttingen University. In 1893 he began work at the Rothschild Museum at Tring, specializing in Coleoptera, Lepidoptera and Siphonaptera. He published over 400 papers, many jointly with Charles and Walter Rothschild. Jordan described 2,575 new species himself, and an additional 851 in collaboration with the Rothschilds. Jordan initiated the first International Entomological Congress. He was a Fellow of the Royal Society, and President of the Entomological Society of

London, 1929-1930.

Biographical Reference: 1960. Riley, N. D. *Biographical Memoirs of Fellows of the Royal Society* **6**:107-133, bibliography, portrait.

Collection

(1) One notebook: *Notes on Sphingidae and on the Literature of Lepidoptera.*
(2) Two Ms notebooks, Tss and drawings of Coleoptera.
(3) Notes and drawings concerning the Anthribidae.
(4) Correspondence collection: approximately 40 letters to and from Jordan concerning the International Commission on Zoological Nomenclature 1914-1927.

198. JURRIAANSE, Jan Hendrick (1867-1940)

Jurriaanse was a Dutch entomological dealer who supplied specimens to individuals and institutions around the world. In addition he possessed an extensive personal collection of butterflies which he gave to the Rijksmuseum van Natuurlijke Historie (Leiden) in 1930. He published *Vogelverdelging. Een woord aan het Nederlandsche Volk* (1914). He was a Fellow of the Nederlandse Entomologische Vereniging, Amsterdam, and the Royal Entomological Society, London.

Biographical Reference: 1986. Smit, P. Hendrik Engel's Alphabetical List of Dutch Zoological Cabinets and Menageries. In *Nieuwe Nederlandse Bijdrajen tot de Geschiedenis der Geneeskunde en der Natuu wetenschappe*, No. 19, 2nd edn, p. 340 (p. 137).

Collection

(1) Three boxes of correspondence consisting of approximately 40 letters to and from Jurriaanse, 1915-1924, with index. Includes letters between Jurriaanse and staff of The Natural History Museum, London and correspondence with James John Joicey (1871-1932) concerning the Hill Museum. Other correspondents include:

Aurivillius, Per Olof Christopher (1853-1928)	1918-1919
Bethune-Baker, George Thomas (1857-1944)	1921-1922
Clark, Benjamin Preston (1860-1939)	1919-1923
Dammerman, Karel Willem (1855-1951)	1921
Eecke, Rudolph van (1886-1975)	1919-1923
Fassl, Anton Heinrich (1876-1922)	1919
Hartert, Ernst (1859-1933)	1918-1923
Jacobson, Edward (1870-1944)	1918-1923
Jordan, Karl Hermann Christian (1875-1972)	1919-1922
Lathy, Percy I. (1870-1943)	1922
Martin, Johannes Karl Ludwig (1851-1942)	1919
Nakamura, Yamoto (1896-1948)	1923
Niepelt, Wilhelm (1862-1936)	1922-1923
Oberthür, Charles (1845-1924)	1919-1922
Oudemans, Johannes Theodorus (1862-1934)	1919-1922
Prout, Louis Beethoven (1864-1943)	1921
Ris, Friedrich (1867-1931)	1922
Roepke, Walter Karl Johann (1882-1961)	1918-1922
Rosenberg, William Frederick Henry (1866-1942)	1919
Schaus, William (1858-1942)	1921
Talbot, George (1882-1952)	1919-1924
Watson, J. Henry (1866-1951)	1919-1922

199. KERREMANS, Charles (1847-1915)

For many years Kerremans was attached to the Belgian military cartographic service, but entomology was his main love. He assembled an important collection of Belgian Coleoptera and later specialized in Buprestidae. He published the multi-volume work *Monographie des Buprestides* (1904-1914), which was unfinished at his death.

Biographical Reference: 1915. Rabaud, E. *Bulletin de la Société Entomologique de France*: 249.

Collection

(1) Correspondence collection: 20 letters, 1903-1904, from Charles Kerremans to Charles Owen Waterhouse (1843-1917) concerning the Kerremans Buprestidae Collection, B.M.1903-59, and one reply from Charles Waterhouse, 1903.
(2) One Ms notebook: [Catalogue of the Kerremans, Buprestid Collection], 1898.

200. KESSEL, Fritz

Collection

(1) One Ts: *Fauna Brasiliensis Coleopterologica* [1932].

201. KEYWOOD, K. P.

In 1948 Keywood presented The Natural History Museum, London with 488 Lepidoptera specimens, 84 Coleoptera and 41 other insects, collected by himself in the Malay Peninsula, 1921-1924. B.M. 1948-135.

Collection

(1) One Ms notebook: *Lepidoptera Collected in the Malay Peninsula 1921-1924.*

202. KIEFFER, Jean Jacques (1856-1925)

A French entomologist, Kieffer taught for many years at the Saint Augustin College, Bitche, Lorraine and was interested in the natural history of the surrounding area. He developed a particular interest in the Hymenoptera and Diptera. His publications included a number of volumes in P. Wytsman's *Genera Insectorum* (1902-1970).

Biographical Reference: 1926. F.W. E[dwards]. *The Entomologist's Monthly Magazine* **62**:44-45.

Collection

(1) One postcard from J.J. Kieffer, postmarked 15 March 1907, addressed to Monsieur Waterhouse, of the British Museum (Natural History).

Insect Room Lists: Volume 1, item 29.

(2) Four postcards from J.J. Kieffer, 1906-1907, addressed to P. Cameron, concerning the Kieffer collections.

Insect Room Lists: Volume 1, item 33.

(3) 3 Ms leaves: List of Kieffer's specimens, 1910.

Insect Room Lists: Volume 2, item 82A.

(4) One postcard from J.J. Kieffer, postmarked 31 October 1909 and addressed to P. Cameron.

Insect Room Lists: Volume 2, item 83(4).

203. KING, James Joseph Francis Xavier (1855-1933)

An artist by profession, King was Professor of Art at the University of Glasgow. He collected British insects, and made frequent collecting trips to the Scottish Highlands, the Lake District, Ireland and the New Forest. He contributed to George Scott's *Fauna, Flora and Geology of the Clyde Area* (1901). King bequeathed his collections to Glasgow University, but some Diptera specimens were later presented by Professor E. Hindle to The Natural History Museum, London. B.M.1938-52.

Biographical Reference: 1933. Anon. *The Entomologist's Monthly Magazine* **69**:166.

Collection

(1) Correspondence collection: Letters between the three entomologists

Prof. E Hindle (1886-1973), Glasgow University, R.L. Coe (1906-1968) and Dr F. W. Edwards (1888-1940) of The Natural History Museum, London concerning the King Diptera collection, 1937-1942.
(2) Ms list: Type specimens from the King collection.

204. KIRBY, William (1759-1850)

Born at Witnesham Hall, Suffolk, Kirby was educated at Cambridge and in 1796 became Rector of Barham near Ipswich, a post he held for 68 years. He published many papers on insects but is best known for his collaboration with William Spence (1783-1860) on *An Introduction to Entomology, or Elements of the Natural History of Insects*, 4 volumes (1815-1826). He was honorary President of the Entomological Society of London and President of the Ipswich Museum.

Biographical Reference: 1852. Freeman J. *Life of W. Kirby, Rector of Barham.* 506pp. London.

Collection

(1) One Ms catalogue: *Museum Entomologicum Barhamaense prima sistens insecta M. Britanniae indigenae.*
(2) Three Ms notebooks: *British Staphylinidae.*
(3) Ms notebook: [Notes on a journey from Barham to Holkham Hall in Norfolk], 1789.

205. KIRBY, William Forsell (1844-1912)

Born in Leicester, Kirby later moved to Brighton, where he became a member of the Brighton and Sussex Natural History Society. In 1867 he became Assistant Naturalist in the Museum of the Royal Dublin Society, and in 1879 he was transferred to the Zoological Department of the British Museum, a post held until his retirement in 1909. He published many important works, including *A Manual of European Butterflies* (1862), *A Synonymic Catalogue of Diurnal Lepidoptera* (1871-1877) and *Synonymic Catalogue of Lepidoptera Heterocera (Moths): Sphinges and Bombyces* (1892). From 1869 to 1884 he contributed to the Lepidoptera part of the *Zoological Record*, and afterwards to a major part of the insect section. He was a Fellow of the Linnean Society, Honorary Secretary of the Entomological Society, 1881-1885, and served on its council in 1886.

Biographical Reference: 1912. Kirby, W.E. *The Entomologist's Record and Journal of Variation* **24**:314-317, portrait.

Collection

(1) Loose-leaf Ms: *On the Genera of Lepidoptera Heterocera established from 1758-1816*. Compiled 1875.
(2) Loose-leaf Ms: *Sketch of the Entomological Collections of the British Museum.*, 1898. Compiled for Lord Walsingham.
(3) One Ms notebook: *Pterophoridae in Universum*, c. 1890.
(4) One Ms notebook: *Tenthredinidae et Uroceridae.*
(5) One Ms notebook: *Notes on Odonata*, c. 1890.
(6) One Ms notebook: *Index to Kirby's [Supplementary] catalogue (Ms) Hesperiidae.*
(7) Six bound volumes: of Ms notes, lists and drawings, c. 1880-1893, including:
  (i) In Volume 3: *Hemiptera-Heteroptera of the Island of Grenada* by P.R. Uhler, c. 1890.
  (ii) In Volume 6: *List of the Hemiptera-Heteroptera collected in the islands of St Vincent by Mr. Herbert H. Smith with descriptions of new genera and species* by P.R. Uhler, c. 1893.
(8) One Ms notebook: Notes by F. Smith.
(9) Loose-leaf Ms: *Bibliography of African Coleoptera*, c. 1890.
(10) One Ms notebook: *Notes and detached descriptions of Orthoptera.*

206. LAFERTÉ SÉNECTÈRE, F. Thibault de la Carte de (1808-1886)

The possessor of a very important collection of Coleoptera, which included specimens from Pierre Dejean (1780-1845), Frederick Bates (1829-1903), François Castelneau (1810-1880) and Ernest Charles Auguste Candèze (1827-1898). He published *Monographie des Anthicus et genres voisins* in 1848.

Biographical Reference: 1886-1887. Anon. *L'Abeille* **25**:174-175, bibliography.

Collection

(1) Two Mss notebooks: *Catalogue des Elatérides communiqués à M. Candèze par M. de la Ferté*, 1857.
(2) Two Ms leaves: [Catalogue of Elateridae sent to Ernest Charles August Candèze (1827-1898) by Laferté Sénectère], 1857.

207. LEECH, John Henry (1862-1900)

Educated at Eton and Trinity Hall, Cambridge, Leech developed an early interest in entomology, and in particular, the Coleoptera and Lepidoptera. He collected insects from Morocco, The Canary Islands, Madeira, Japan, Korea, China and the north-western Himalayas, but after 1887 stayed in Britain to work on his collections, and commissioned others to collect for him. Leech published *British Pyralides* (1886) and *Butterflies from China, Japan and Corea*, 3 vols (1892-1894). In 1889 he purchased the entomological periodical *The Entomologist*, partly as a means of publishing papers on his own collections. He withdrew from the journal after objections were made about the inclusion of these publications. He was a Fellow of the Linnean Society and of the Entomological Society of London. He was also a member of the Société Entomologique de France, and of the Entomologischen Verein zu Berlin.

Biographical Reference: 1901. Anon. *The Entomologist* **34**:33-38, bibliography, portrait.

Collection

(1) One Ms notebook: *Leech's Butterflies from China, Japan and Corea.* Contains species information as given in the published work of the same title.

208. LEEDS, Henry Attfield (1873-1958)

Born at Somersham, Huntingdonshire, he worked on the railways until he retired in 1933 and devoted the rest of his life to the study of British Lepidoptera. With P.M. Bright (1865-1941) he wrote *A Monograph of the British*

Aberrations of the Chalk Hill Blue Butterfly Lysandra coridon (*Poda, 1761*) (1938), and with L.W. Newman (1873-1949), *Text Book of British Butterflies and Moths* (1913). He was a member of the South London Entomological Society.

Biographical Reference: 1959. Anon. *The Entomologist* **92**:22.

Collection

(1) Correspondence collection: 11 letters, 1936-1941, between H.A. Leeds and N.D. Riley (1890-1979), Keeper of Entomology at The Natural History Museum, London regarding the addenda to *A Monograph of the British Aberrations of the Chalk Hill Blue Butterfly* Lysandra coridon.
(2) Two boxes: Coloured plates of British Rhopalocera.
(3) One loose-leaf file: *Aberrations of Female Uppersides of the Chalk-hill Blue Butterfly*, 1938, with hand-coloured half-tone plates, notes and index.

209. LE MASURIER, P.

Collection

(1) Two Ms diaries: *Rhopalocera and Heterocera*, 1945- 1961.

210. LORT- PHILLIPS, Frederick Alfred (1872-1944)

Educated at Eton and at Trinity College, Cambridge. He was a traveller and sportsman who shot big game in many parts of the world. He was particularly interested in mammals and birds, but in 1931 he presented 3,600 Lepidoptera specimens collected by him from various British localities to The Natural History Museum, London. B.M.1931-459. He was Vice-President of the Zoological Society of London.

Biographical Reference: 1944. WLS. *Ibis* **86**:230.

Collection

(1) Two Ms diaries: Catalogue of British Lepidoptera collection, B.M.1931-459.

211. LUCAS, William John (1858-1932)

Educated at a grammar school in Oxford and at the University of London, Lucas became a teacher at Tiffin Boys School, Kingston-on-Thames, and was appointed Lecturer in Nature Study to the Surrey County Council. Interested in all aspects of natural history, especially entomology, he became the chief authority on British Orthoptera, Odonata and Neuroptera. Lucas's publications include *British Dragonflies* (1900), *British Orthoptera*. (1917) and *The Aquatic (Naiad) Stage of the British Dragonflies* (1930). In 1901 he joined the editorial panel of *The Entomologist*. He served on the council of the Entomological Society, and was President of the South London Entomological Society, and Vice-President of the Lancashire and Cheshire Entomological Society.

Biographical Reference: 1932. Killington, F.J. *The Entomologist* **65**:25-27, bibliography, portrait.

Collection

(1) Ten Ms notebooks: *Natural History Notes*, 1887-1931.

212. MABILLE, Paul (1835-1923)

Both a botanist and entomologist, Mabille published a number of Lepidoptera papers, and with Paul Vuillot, *Novitates Lepidopterologicae* (1890-1895). He was President of Société Entomologique de France.

Biographical Reference: 1923. Anon. *Entomological News* **34**:256.

Collection

(1) One loose-leaf Ms: [Descriptions and drawings of Rhopalocera, including species from Africa.]
(2) One loose-leaf Ms: *Mabille's Original American Descriptions and Figures* [of Lepidoptera as arranged by Charles Oberthür], 1845-1924.

213. MACFIE, John William Scott (1879-1948)

Born in New Ferry, Cheshire, Macfie studied at Cambridge University. In 1910 he joined the West African Medical Staff and remained in Africa until his retirement in 1923. He carried out pioneering work in the study of mosquitoes and medical entomology. During World War II Macfie worked at a field malaria laboratory in the Near East, and later joined the Liverpool School of Tropical Medicine, publishing papers on West African tsetse flies and *Aedes aegypti*. Macfie later studied ceratopogonid midges and published a series of papers (1921-1924) on Ghanaian species.

Biographical Reference: 1948. Mattingly, P.F. *The Entomologist* **82**:21-22.

Collection

(1) One Ts: *New species of the genera* Forcipomyia *and* Atrichopogon *from the State of Chiapas* [Mexico].
(2) One Ts: [Catalogue of Ceratopogonidae], 1884-1948.
(3) One copy of a letter, 1946, from Alfonso Dampf to Macfie, concerning the shipment of Mexican Ceratopognidae, with a list of localities. Collection 19,660 specimens. B.M.1950-452
(4) Correspondence collection: 14 letters, 1947-1948, from Fritz Isidore van Emden (1898-1958), to Macfie.

214. McKAY, Harold

McKay was a keen butterfly collector. His fine collection of Rhopalocera from Malawi, consisting of over 1,300 specimens was presented to The Natural History Museum, London in 1980. B.M.1980-574.

Collection

(1) One box, containing:
    (i) One letter to McKay from David Gifford, 1960.
    (ii) One *Setting Book*, 1952-1964.
    (iii) Index cards to collection.

215. McLACHLAN, Robert (1837-1904)

McLachlan's first interest in entomology was the Lepidoptera, but he soon concentrated on Trichoptera and Odonata. In 1870, with assistance from A.E. Eaton, he published, *Catalogue of the British Neuroptera*, and between 1874 and 1884, *A Monographic Revision and Synopsis of the Trichoptera of the European Fauna*. He was a pioneer in the study of the fauna of the French and Swiss mountains, exploring the Dauphiny and Savoy Alps. The Belgian entomologist, Baron de Selys Longchamps, became a close friend and they collaborated on many projects. McLachlan contributed many papers to the *Entomologist's Weekly Intelligencer* and helped in the launch and continuation of *The Entomologist's Monthly Magazine*. He was a Fellow of the Entomological Society of London and served as President, Secretary and Treasurer. He was also a Fellow of the Royal Society, the Linnean Society, the Royal Horticultural Society and a member of many foreign entomological societies. His large insect collection was purchased by The Natural History Museum, London in 1938 and contained over 50,000 specimens including 12,750 Odonata, 14,000 Trichoptera and 500 types.

Biographical Reference: 1904. Eaton, A.E. *et al. The Entomologist's Monthly Magazine.* **40**:145-148.

Collection

(1) One folder, containing:
 (i) Correspondence collection: Letters between Charles McLachlan and the Entomology Department, The Natural History Museum, London c. 1927-1938, regarding the purchase of the McLachlan collection.
 (ii) Correspondence collection including letters from:

Durrant, J.H.	1903-1904
McLachlan, Hugh	1904
McLachlan, Robert	1904

 (iii) Portrait of Robert McLachlan.

(2) One Ms notebook: [Descriptions of Odonata taxa], also miscellaneous correspondence, c. 1874.

216. MALARIA MOSQUITOES

Miscellaneous documents on collecting malaria mosquitoes.

Collection

(1) One Ms volume: entitled *Malaria Mosquitoes* includes *The First Preliminary Report upon the Culicidae of the Gold Coast Colony, West Africa* by Albert J Chalmers, 1899.

217. MANNING, Francis Joseph (1912-1966)

A lecturer in biology at the Cheshire College of Education, his particular interest was bee-keeping and bees in general. He was a Fellow of the Royal Entomological Society.

Biographical Reference: 1967. Pearson, E.O. *Proceedings of the Royal Entomological Society of London* (C) **31**:62.

Collection

(1) One Ts: *Sex Determination in* Apis mellifera L.
(2) One Ts: *A Report on a Cytological Investigation of some Bees and Wasps (Hymenoptera) with Special Reference to their Sex-determining Mechanism.*

218. MARSH, Dudley Graham (1891-1969)

During his early life he was a keen butterfly collector, later his interests changed to the study of moths.

Biographical Reference: 1969. G.H.Y. *Entomologist's Record and Journal of Variation* **81**:312.

Collection

(1) Two Ms notebooks: [Details of Lepidoptera specimens in the Marsh Collection].

219. MARSHALL, Guy Anstruther Knox (1871-1959)

Born in India and educated at Charterhouse School, Marshall became interested in entomology from an early age. His particular interest was the Coleoptera, especially Curculionidae. He published some 200 works on their systematics, including the Curculionidae volume in *The Fauna of British India*, edited by W.T.Blanford. Marshall was responsible for the establishment of the *Bulletin of Entomological Research* in 1909, and the *Review of Applied Entomology* in 1913. He was the first Director of the Commonwealth Institute of Entomology. He was elected an Honorary Fellow of the Royal Entomological Society of London.

Biographical Reference: 1959. Anon. *Annals and Magazine of Natural History* (13) **1**:753-754, portrait.

Collection

(1) One loose-leaf Ms: [Data on the capture of South African Coleoptera, 1891-1898].

220. MARSHALL, Thomas Ansell (1827-1903)

Born at Keswick, Marshall studied Classics at Oxford University. Later he took holy orders, holding a number of livings, including a period in Antigua in the West Indies. On his retirement he went to Corsica. Although interested in all aspects of entomology, he had a special interest in parasitic Hymenoptera. Important publications include *Ichneumonidum Britannicorum Catalogus* (1870), *A Catalogue of British Hymenoptera; Chrysididae, Ichneumonidae, etc.* (1872) and *A Catalogue of British Hymenoptera Oxyura* etc. (1873) and the Braconidae in André's *Species des Hyménoptères d'Europe*, (1879-1913). Marshall was a Fellow of the Entomological Society of London and of the Linnean Society.

Biographical Reference: 1903. McLachlan, R. *The Entomologist's Monthly Magazine* **39**:152-153.

Collection

(1) Five Ms leaves: *The Rev. T.A. Marshall's Collection of Hymenoptera*

Insect Room Lists: Volume 2, item 85.

221. MARSHAM, Thomas (d. 1819)

A well educated man, he was secretary to the West India Dock Company and an officer in the Volunteer Corps. He developed a special interest in the Coleoptera and in 1802 he published *Coleoptera Britannica,* Volume 1 of *Entomologia Britannica.* His collection of Coleoptera was purchased by J.F. Stephens (1792-1852) and was incorporated into his own collection, now in The Natural History Museum London, B.M.1853-46.

Biographical References: 1920, Fletcher, J.J. *Proceedings of the Linnean Society of New South Wales* **45**:570-571, 574; 1976. Mackechnie Jarvis, C. *Proceedings of the British Entomological and Natural History Society* **8**:92.

Collection

(1) One Ms: [Notes by G.R. Waterhouse (1810-1888) on Marsham's and Stephens' type specimens.] B.M.1853-46.

Insect Room Lists: Volume 2, item 92.

(2) Two Ms leaves: [List of specimens with sale prices, the specimens originally from the Entomological Society, were purchased from Marsham by Stephens.]

Insect Room Lists: Volume 2, item 92.

(3) Nine Ms leaves: [Insects from Marsham's cabinet and a catalogue of British Insects] 1820.

Insect Room Lists: Volume 1, item 37.

222. MASON, Philip Brookes (1842-1903)

Born in Burton-on-Trent, Mason studied medicine at Glasgow University. Outside of medicine, he developed an interest in entomology. His special interests included the Coleoptera, but he studied most aspects of British natural history. During his lifetime he acquired many important collections, including that of E.C. Rye, A. Matthews and the Aculeate Hymenoptera of F. Smith. Mason was a Fellow of the Linnean Society and the Entomological Society of London.

Biographical Reference: 1904. Fowler, W.W. *The Entomologist's Monthly Magazine* **40**:17-18.

Collection

(1) One Ts letter from Mason to C.O. Waterhouse dated 1899.

Insect Room Lists: Volume 2, item 1.

223. MASTERS, George (1837-1912)

Born in Kent, Masters went to Australia in 1856 and worked for Dr. Godfrey Howitt (1800-1873) who was a keen entomologist. In 1859 he met William Macleay and collected for him. In 1864 Masters was appointed Assistant Curator at the Australian Museum. In 1871-1872 he published, *Catalogue of the Described Coleoptera of Australia,* in four parts.

Biographical Reference: 1983 Highley, E. and Taylor, R.W. *Australian Systematic Entomology : A Bicentenary Perspective.* Melbourne: Commonwealth Scientific and Industrial Research Organization, 147pp. (pp.12-13), portrait.

Collection

(1) One Ms notebook: Ms *index to the Catalogue of the Described Coleoptera of Australia by George Masters 1871-72. [By Gilbert Ernest Bryant (1878-1965).]*

224. MAULIK, Samarendra (1881-1950)

Born in Tamluk, India, he studied physics, chemistry and mathematics at St Xavier's College, Calcutta. In 1911 Maulik travelled to England and studied zoology at Cambridge and Imperial College, London. Later he joined the staff of The Natural History Museum, London and remained there for most of his working life, studying the Chrysomelidae. In 1919 he published the first of three volumes on the Chrysomelidae in the series *Fauna of British India*; the second and third volumes were published in 1926 and 1936.

Biographical Reference: 1950. Bairstow, L. *The Entomologist's Monthly Magazine* **86**:225-226.

Collection

(1) One bound Ts: *An Inquiry into the Relationships between the Chrysomelid Beetles and their Food-plants*, 1945.
(2) Three work diaries, 1934-1947.
(3) Mss and drawings of Chrysomelids.
(4) Correspondence collection: approximately 300 letters to Maulik from various entomologists, c. 1929-1948.

225. MAXWELL, Reginald Maitland (d. 1967)

Maxwell served in the Indian Civil Service from 1906 to 1944, becoming Home Secretary and Home Member of the Viceroy's Council. During this time he made extensive collections of Indian Lepidoptera. Over a number of years his collection of 6,000 Lepidoptera came to The Natural History Museum, London. B.M.1950-397; 1950-398; 1951-22; 1967-553.

Collection

(1) One typed Ms : *History of the Maxwell Collection of Lepidoptera* by Lyle Maxwell, c. 1975. Includes a list of new species named and described by Edward Meyrick in the Maxwell Collection.
(2) Photographic collection: [Early stages of Indian Lepidoptera], 1911-1938.

226. MEDENBACH, Alexander Benjamin de Roöy van (1841-1878)

His particular entomological interests concerned the European and exotic Lepidoptera, but he was also interested in the Hemiptera and Coleoptera of the Netherlands. Medenbach died in Arnhem. He was a member of the Netherlands Entomological Society.

Biographical Reference: 1878. Kraatz, G. *Deutsche Entomologische Zeitschrift* **22**:226.

Collection

(1) One Ms notebook:
  (i) *Catalogue of Hocking Collection.*
  (ii) *Californian Noctuae. New Species*, c. 1875.

227. MEEK, E. G.

Collection

(1) Loose-leaf Ms: *Notes on the* Sesia n.d. [a paper read to members of the West London Entomological Society].

228. MEIGEN, Johann Wilhelm (1764-1845)

Born in Solingen, Germany, he lived in Stolberg near Aachen. Meigen had a wide interest in natural history but his particular area of study was the classification of Diptera. His important publications include: *Klassifikazion und Beschreibung der europäischen zweiflügeligen Insekten* (1804), and *Systematische Beschreibung der bekannten europäischen zweiflügeligen Insekten* (1818-30).

Biographical Reference: 1846. Forster, J.A. *Stettiner Entomologische Zeitung* **7**:66-74, 130-141.

Collection

(1) One Ms notebook containing:
  (i) *Abbildungen zweiflügel. Insekten nach der Natur, Tab. 1-305, in 2 Quartbände (dem Pariser Museum verkauft).*

(ii) *Verzeichniss der von Ritter v. Linné beschriebenen zweiflügeligen Insekten.*

(iii) *Verzeichniss der in meinen Werken beschriebenen zweiflügeligen Insekten.*

(iv) *Verzeichniss der von Prof. [J.W.] Zetterstedt in Lappland neu entdeckten zweiflügeligen Insekten.*

(v) *Übersicht der bis 1840 bekannten* [zweifl] *Insekten nach den Welttheilen geordnet.*

(2) One notebook: [Original pen-and-ink drawings of the heads and mouthparts of Diptera, with index.]

(3) One Ms: *Systema Dipterorum Europaearum* [1835].

(4) Collection of manuscripts and drawings including:

(i) Plates from *Klassifikazion und Beschreibung der europäischen zweiflügeligen Insekten* (1804). Handcoloured by the author.

(ii) One copy of *Abbildung aller bis jetzt bekannten europäischen zweiflügeligen Insekten,* (1830). A second copy, with coloured plates and Ms list of names. A third copy, with uncoloured plates and Ms list of names.

(iii) Notes and criticisms received from [J.C.] Graf von Hoffmansegg with Meigen's signed draft reply, 1814-1815.

(iv) Original pen-and-ink drawings of Diptera parts.

(v) 35 sheets of original watercolour drawings of Diptera.

vi) *Ausländische zweiflügelige Insekten.* Ten sheets of original watercolour drawings.

vii) Eight sheets of pen-and-ink drawings of Diptera.

(5) One notebook: Synopsis of known Diptera [1806-17].

229. MEYRICK, Edward (1854-1938)

Born at Ramsbury, Wiltshire, Meyrick studied at Trinity College, Cambridge. In 1877 he went to Sydney, Australia, and later to Christchurch, New Zealand. In 1887 he returned to England, becoming a teacher at Marlborough School. He developed an early interest in the Microlepidoptera and published many papers on their classification. Publications include, *A Handbook of the British Lepidoptera* (1895), *A Revised Handbook of British Lepidoptera* (1928), and contributions to Wytsman's *Genera Insectorum* (1902-1970). During his life he described over 20,000 new species of Lepidoptera and hundreds of new genera. His collection of 100,000 specimens was bequeathed to The Natural History Museum, London. B.M.1938-290. He was elected a Fellow of the Royal Society and a Fellow of the Royal Entomological Society, London.

Biographical Reference: 1939. Hall, A.W. *Obituary Notices of Fellows of the Royal Society* **2**:531-548, bibliography, portrait.

Collection

(1) Approximately 100 watercolour drawings of Microlepidoptera copied from R. Felder and A.F. Rogenhofer, *Reise der Österreichischen Fregatte Novara*, Zoologischer Theil, Bd. II. Lepidoptera (1864-1875).
(2) One Ms volume: [Notes on the Microlepidoptera].
(3) 32 volumes of Ms relating to the Microlepidoptera collection, B.M.1938-290.

(i)	*Index to genera and species in catalogues 6,7,10.*
(ii)	*Index to Pyrales.*
(iii)	*Indexed list of South African Microlepidoptera.*
(iv)	*Indexed list of Asiatic Microlepidoptera.*
(v)	*Indexed list of Australian Pyrales and Microlepidoptera.*
(vi)	*Catalogue of Tineina, Hepialina (Genera and species).*
(vii)	*Catalogue of Tineina (Genera and species).*
(viii)	*Indexed list & descriptions of Pyralidae genera.*
(ix)	*An early list of micros taken by Meyrick and arranged after Staudinger and Wocke, c. 1875.*
(x)	*Catalogue of Tortricina (Genera and species).*
(xi)	*A list of Pyralid Genera with type species & original references.*
(xii)	*Food plants.*
(xiii)	*Catalogue of Pyralidina (Genera and species) with index to Genera.*
(xiv)	*Generic descriptions of Tineina.*
(xv)	*List of micro genera.*
(xvi)	*Generic descriptions of Tortricina.*
(xvii)	*Notes on larvae & pupae.*
(xviii)	*Copied descriptions.*
(xix-xxiv)	*Determinations. 1897-1938.*
(xxv)	*Meyrick's diary of Captures, 1872-1890.*

(xxvi) *Copied descriptions of Geometridae and Pyralidae.*

(xxvii-xxix) *Copied descriptions of genera of microlepidoptera.*

(xxx) *Original descriptions of Microlepidoptera.*

(xxxa) *Descriptions of World Genera with keys and index.*

(xxxi) *Species of Tineina, Europe.*

(4) One Ms notebook: *List of Coleoptera (British). Diary of captures Australia 1877-1878.*

(5) Three Ms notebooks: *Meyrick's notes on Revised Handbook British Lepidoptera*, 1928.

(6) One notebook: *Numerical index to negatives of genitalia – photomicrographs – giving slide no. Mostly Meyrick species of Microlepidoptera taken by H. M. Malies.*

(7) One Ms notebook: *Sketches of Tineid venation.*

(8) One Ms notebook: [Systematic catalogue.]

(9) One Ms notebook: *List of Meyrick's Palaearctic Coll. with index and list of types taken from the old British Coll.*

(10) One Ms notebook: *Meyrick's key to Genera of Phycitinae.*

230. MILES–MOSS, Arthur (1873-1948)

The Rev. Arthur Miles-Moss was a student at Cambridge and then, in 1897, travelled to South America. From 1907 to 1910 he lived in Peru, and from 1911 to 1945 he stayed at Pará in Brazil. He made large collections of insects, but was particularly interested in the Lepidoptera. He bred many species and made drawings of larvae and pupae. He was a talented artist, painting many land- and seascapes and tree studies. Moss was an exhibitor at the Royal Academy. In 1947 The Natural History Museum, London purchased his collection of 25,000 Brazilian Lepidoptera. B.M.1947-453.

Biographical References: 1926. Anon. *Westmorland Gazette*, 6 November, 1926; 1948. Williams, C.B. *Proceedings of the Royal Entomological Society of London* (C) **12**:65.

Collection

(1) Two bound volumes: [Annotated watercolour drawings of Lepidoptera larvae drawn in Pará, Brazil], c. 1929-1935.

(2) One bound volume: [Contains a copy of Miles-Moss, A., Sphingidae of Pará, Brazil, *Novitates Zoologicae,* Vol. 27 (1920), pp. 333-424, with annotations and original drawings.]

(3) Four boxes of Mss: [Including notes on the Lepidoptera and birds of the Amazon, the church, journeys in the Amazon and copies of letters] c. 1946.

(4) Correspondence relating to the purchase of the Miles-Moss collection by The Natural History Museum, London, c. 1946.

(5) Miscellaneous collection, includes drawings, photographs, notes and letters, 1933-1945.

(6) Correspondence collection: approximately 300 letters, 1898-1945, includes letters from:

Boy, Hugo Carlos (d. 1937)	1935
Eltringham, Harry (1873-1941)	1927-1932
Evans, William Harry (1876-1956)	1937
Fountaine, Margaret Elizabeth (1862-1940)	1930-1932
Jordan, Karl Hermann Christian (1875-1972)	1908-1932
Kaye, William James (1875-1967)	1911-1936
Littlewood, Frank (1882-1949)	1926
Parish, Herbert Simpson (1870-1957)	1926
Poulton, Edward Bagnell (1856-1943)	1926-1937
Querci, Orazio (1875-1967)	1920
Rothschild, Nathaniel Charles (1877-1923)	1898-1917
South, Richard (1846-1932)	1900-1905
Talbot, George (1882-1952)	1916
Townsend, Charles Henry Tyler (1863-1944)	1926
Tutt, James William (1858-1911)	1903-1907
Wallace, Alfred Russel (1823-1913)	1909-1910
Zerny, Hans (1887-1945)	1927

(7) Correspondence collection: approximately 100 letters, 1912-1921, from H.J. Thouless to Miles-Moss.

231. MILLER, [H].

Collection

(1) One Ms notebook: [Reminiscences of collecting Lepidoptera around London], c. 1920s.

232. MNISZECH, Georges Vandalin von (d. 1881)

Born in Poland, Mniszech came from a wealthy family. He developed interests in a range of subjects including geology, palaeontology, astronomy and the arts. However, entomology was his main interest and he formed a large Coleoptera collection, which passed to the Muséum Nationale d'Histoire Naturelle, Paris.

Biographical Reference: 1881. Deyrolle, H. *Bulletin de la Société Entomologique de France* (6) **1**:cxlii-cxliv.

Collection

(1) Manuscript notebook: *Catalogue Buprestidae de la collection de M. le Comte de Mniszech,* 1868.

233. MOCSÀRY, Alexander (Sándor) (1841-1915)

Born in Nagyvarad, Mocsàry became a leading Hungarian entomologist with special interest in the Chrysidoidea. He worked for 44 years in the National Museum, Budapest and published over 150 papers.

Biographical Reference: 1922. Viereck, H.L. *Entomological News* **33**:157-158.

Collection

(1) Six Ms leaves: *List of Hymenoptera (Chrysidoidea) determined by Dr. A. Mocsàry, 1913.*

Insect Room Lists: Volume 2, item 107.

234. MOORE, Frederic (1830-1907)

Moore joined the staff of the Museum of the East India Company in 1848 and retained the position of Assistant Curator until 1879, when the insect collection was transferred to the British Museum. He developed an extensive collection of butterflies and moths and was known as 'the father of Indian Lepidopterists'. His publications include *A Catalogue of the Lepidopterous Insects in the Museum of the Honourable East India Company* (1857-1859), co-authored with T. Horsfield; *Descriptions of New Indian Lepidopterous Insects (Heterocera) from the Collection of the Late W.S. Atkinson* 1879-88; and the *Lepidoptera of Ceylon* (1880-1887). In 1890 the first part of the 10-volume work *Lepidoptera Indica* was published and was completed after Moore's death by C. Swinhoe (1836-1923). Moore was a Fellow of the Entomological Society of London.

Biographical Reference: 1907. Anon.*The Entomologist's Monthly Magazine* **43**:162.

Collection

(1) Four boxes of Ms notes: [Indian Coleoptera], c. 1880.
(2) Two boxes of Ms notes: *Monographic Revision of the Genera of European and Asiatic Satyrinae*, c. 1897.
(3) Ten boxes of notes, diagrams and original drawings relating to *Lepidoptera Indica* (1890-1913), including some letters from A.G. Butler (1844-1925), 1899.
(4) Correspondence collection:

McLachlan, Robert (1837-1904)	1875
Marshall, Guy Anstruther Knox (1871-1959)	1870
Oberthür, Charles (1845-1924)	1883-1891
Westwood, John Obadiah (1805-1893)	1875-1880

235. MORLEY, Claude (1874-1951)

As a boy, Morley developed a keen interest in butterflies, but later concentrated on the Hymenoptera, publishing many papers on the British parasitic

Hymenoptera. In 1903 he commenced publication of *Ichneumonologia Britannica.* In 1909 he joined The Natural History Museum, London and in 1910 published the *Catalogue of British Hymenoptera of the family Chalcididae* and between 1912 and 1915 published *A revision of the Ichneumonidae.* He lived for many years in Suffolk and in 1929 founded the Suffolk Naturalists' Society.

Biographical References: 1951. Anon. *Transactions of the Suffolk Naturalists' Society* **7**:lxxvii-lxxviii, portrait; 1951. Riley, N.D. *The Entomologist* **85**:121-122.

Collection

(1) One loose-leaf typed Ms: *Catalogus Musei Mei.* List of insects, plants and minerals in Morley's natural history collection, 1946.
(2) One Ms notebook: *List of my Entomological Collection, Dec 31st 1895.*
(3) Album of signatures of nineteenth- and twentieth-century entomologists:
Adams, Frederick Charlstrom (d. 1920)
Adkin, Robert (1849-1935)
André, Jacques Ernest Edmond (1844-1891)
Andrews, Henry William (1876-1955)
Bailey, James Harold (1870-1909)
Baker, William A. (1899-1973)
Balfour-Browne, William Alexander Francis (1865-1958)
Bankes, Eustace Ralph (1861-1929)
Barker, Henry William (1860-1909)
Barrett, Charles Golding (1836-1904)
Bates, Frederick (1829-1903)
Bayford, Edwin Goldthorp (1865-1958)
Beare, Thomas Hudson (1859-1940)
Beaumont, Alfred (1832-1905)
Bedwell, Ernest Charles (1875-1945)
Bennett, William Henry (1862-1931)
Bignell,George Carter (1826-1910)
Bingham, Charles Thomas (1848-1908)
Blair, Kenneth Gloyne (1882-1952)
Bloomfield, Edwin Newson (1827-1914)

Brunetti, Enrico Adelelmo (1862-1927)
Butler, Edward Albert (1845-1925)
Butterfield, Rosse (1875-1935)
Cameron, Peter (1847-1912)
Campion, Herbert (1870-1924)
Carpenter, Geoffrey Douglas Hale (1882-1953)
Carr, John Wesley (1862-1939)
Champion, George Charles (1851-1927)
Chaster, George William (d. 1910)
Chitty, Arthur, John (1859-1908)
Collet, Edward Pyemont (1863-1937)
Collin, James Edward (1876-1968)
Cottam, Arthur (1837-1911)
Cummings, Bruce Frederic
(pseudonym: Wilhelm Nero Pilate Barbellion) (1899-1919)
Curtis, William Parkinson, (1878-1968)
Dale, Charles William (1851-1906)
Davies, Edwin C. H. (d. 1909)
Day, Frank H. (1875-1963)
Donisthorpe, Horace St. John Kelly (1870-1951)
Douglas, John William (1814-1905)
Durrant, John Hartley (1863-1928)
Edelsten, Hubert MacDonald (1877-1959)
Edwards, James (1856-1928)
Edwards, Stanley (1864-1938)
Elliott, Ernest A. (d. 1936)
Evans, William Harry (1876-1956)
Fitch, Edward Arthur (1854-1912)
Fletcher, John Edward (1836-1902)
Fordham, William John (1882-1942)
Froggatt, Walter Wilson (1858-1937)
Gahan, Charles Joseph (1862-1939)
Garde, Philip Hardy de la (d. 1913)
Gibbs, Arthur Ernest (1859-1917)
Gorham, Henry Stephen (1839-1920)

Gosse, Herbert (d. 1908)
Green, Edward Ernest (1861-1949)
Grimshaw, Percy Hall (1869-1939)
Haines, Frederick Haslewood (1864-1946)
Halbert, James Nathaniel (1872-1948)
Hamm, Albert Harry (1861-1951)
Harrison, Albert (1860-1911)
Harwood, Bernard Smith (1876-1933)
Hewitt, Charles Gordon (1885-1920)
Howard, Leland Ossian (1857-1950)
Image, Selwyn (1849-1930)
Jacob, John Kenneth (d. 1941)
Janson, Oliver Erichson (1850-1926)
Jenkinson, Francis John Henry (1853-1923)
Jones, Albert Hugh (1840-1924)
Jordan, Karl Hermann Christian (1875-1972)
Joy, Norman Humbert (1874-1953)
Keys, James H (1855-1941)
Killington, Frederick James (1894-1956)
King, James Joseph Francis Xavier (1855-1933)
Kirby, William Forsell (1844-1912)
Kirkaldy, George Willis (1873-1910)
Knaggs, Henry Guard (1832-1908)
Lefroy, Harold Maxwell (1877-1925)
Lucas, William John (1858-1932)
Luff, William Ambridge (1851-1910)
Lyle, George Trevor (1873-1930)
Main, Hugh (d. 1948)
Malloch, John Russell (1874-1963)
Marshall, Guy Anstruther Knox (1871-1963)
Marshall, John Frederick (1874-1959)
Meyrick, Edward (1854-1938)
Morey, Frank (1858-1925)
Morley, Arthur Macdonell (1879-1972)
Nevinson, Edward Bonney (1858-1927)

Newberry, Emanuel Augustus (1845-1927)
Nurse, Charles George (1862-1933)
Nurse, Euston John (1865-1945)
Ormerod, Eleanor Anne (1828-1901)
Oshanin, Vasilii Fedorovich (1844-1917)
Pickard-Cambridge, Octavius (1828-1917)
Porritt, George Taylor (1848-1927)
Power, John Arthur (1810-1886)
Prout, Louis Beethoven (1864-1943)
Rait-Smith, William (1875-1958)
Ransom, Edward (d. 1946)
Richardson, Nelson Moore (1855-1925)
Roebuck, William Denison (1851-1919)
Rothney, George Alexander James (1849-1922)
Rothschild, Nathaniel Charles (1877-1923)
Routledge, George Bell (1864-1934)
Saunders, Edward (1848-1910)
Sharp, David (1840-1922)
Sharp, William E (1856-1919)
Sich, Alfred (d. 1943)
Sladen, Frederick William Lambert (1876-1921)
Stephens, James Alfred (1882-1947)
Stevens, Samuel (1817-1899)
Swinton, Archibald Henry (1845?-1936)
Taylor, John Kidson (1839-1922)
Thornley, Alfred (1855-1947)
Tutt, James William (1858-1911)
Verrall, George Henry (1848-1911)
Walker, James John (1851-1939)
Waterhouse, Charles Owen (1843-1917)
Waterhouse, Edward A (1851-1916)
Waterston, James (1879-1930)
Whittingham, Walter Godfrey (1861-1941)
Whittle, Francis George (1854-1921)
Wickham, Archdale Palmer (1855-1935)

Woodforde, Francis Cardew (1846-1928)

Yerbury, John William (1847-1927)

(4) Author's annotated copies of *The Hemiptera of Suffolk* (1905) and *The Hymenoptera of Suffolk*, Part 1: *Aculeata*, (1899).

(5) One loose-leaf Ts: *Oxyura Britannica: The Proctotrupidae of Great Britain*, 1917.

(6) One loose-leaf Ts : *Familiar Gnats and Flies of the British Isles.*

(7) One loose-leaf Ts : *Catalogus Ichneumonidum Africae,* 1917.

(8) One letter from Thomas Ansell Marshall (1827-1903) to C. Morley dated 1899.

(9) One loose-leaf Ms: *A Synopsis of British Proctotrupidae (Oxyura)*, 1922-1923.

236. MORRIS, D.

Collection

(1) Two Ms leaves: *List of insects sent to be named by Mr Morris, Barbados*. B.M. 1902-5.

Insect Room Lists: Volume l, item 24.

237. MORRIS, Stanley (*fl.* 1877-1942)

Morris had a wide interest in natural history, but was particularly interested in butterflies and birds. Living in West Sussex he kept detailed diaries on natural history and observations in the Sussex area.

Collection

(1) 40 natural history diaries, 1889-1942 covering the Sussex area, and one index volume.

(2) Three bird notebooks:

  (i) *A Reference to the Occurrences of the Rarer Species of British Birds in Sussex Collected from Various Sources.*

  (ii) *Immigration Charts of the Summer Residents as recorded in Mid & West Sussex between 1890 and 1925 inclusive.*

(iii) *Bird Song: Its Import and Significance.*

(3) Two notebooks: *Rough Notes on British Butterflies recorded in West Sussex,* c. 1930s.

(4) Three Ms notebooks: *Short notes on natural history, c. 1900.*

(5) One Ms notebook: *A book of records for the county of Sussex, and more particularly of local records taken from my diary of observations and occurrences in natural history for Lewes and neighbourhood.*

238. MOSELY, Martin Ephraim (1867-1948)

A keen angler, Mosely became interested in the study of Trichoptera, and to a lesser extent the Ephemeroptera and Plecoptera, in relation to fly-fishing. He collected Trichoptera in Switzerland, France and Corsica, and on his retirement in 1929 he studied Trichoptera at The Natural History Museum, London. He published over 100 papers on entomology and many on angling subjects. Publications include: *The Dry-fly Fisherman's Entomology* (1921); *Insect Life and the Management of a Trout Fishery* (1926); and *The British Caddis Flies* (1939). He bequeathed his collection of over 7,000 Trichoptera specimens to The Natural History Museum, London. B.M.1948-589.

Biographical Reference: 1948. Kimmins, D.E. *The Entomologist's Monthly Magazine* **84**:240.

Collection

(1) Four photograph albums, mostly Trichoptera.

(2) One Ms notebook: [Notes on collecting Trichoptera and fishing], 1907-1910.

239. MUIR, Frederick Arthur Godfrey (1872-1931)

Born in Clapham, London, Muir worked for the Eastern Telegraph Company in Aden, Mozambique and Durban. He collected specimens and sent many to The Natural History Museum, London. In 1905 he was appointed assistant entomologist to the Hawaiian Sugar Planters' Association, working under R.C.L. Perkins. His work concentrated on introducing parasites to control insect pests of sugar-cane. This involved travel in the Malay States, Borneo, Java and the Moluccas. He published over 100 papers including work on the Fulgoroidea. He was a Fellow of the Entomological Society of London and was twice President of the Hawaiian Entomological Society.

Biographical References: 1931. Swezey, O.H. *Hawaiian Planters' Record* **35**:205-215, bibliography; 1932. Williams, F.X. *Proceedings of the Hawaiian Entomological Society* **8**:13-15, 141-152, bibliography, portrait.

Collection

(1) Two boxes of manuscripts including:
  (i) *Description of genera of Cixiidae and Achilidae.*
  (ii) *On the classification of the Fulgoroidea.*
  (iii) *Cixiid notes.*
  (iv) *Terminal abdominal segments and genitalia.*
  (v) Miscellaneous drawings of Homoptera.
  (vi) Miscellaneous notes.
(2) Two Ms notebooks: *Delphacidae.*
(3) Two notebooks: *Cixiidae.*

240. MUMFORD, Edward Philpott and ADAMSON, Alistair Martin (1901-1945)

In 1928 Adamson joined the Pacific Entomological Survey, under the Directorship of a fellow Englishman, Edward Mumford. The Survey was organized and financed by the Hawaiian Sugar Planters' Association and the Bernice P. Bishop Museum in Honolulu to study the fauna of the Pacific Islands. Adamson and Mumford spent almost two years in the Marquesas Islands, where they made extensive collections of invertebrates.

Biographical Reference: (Adamson) 1946. Anon. *Tropical Agriculture* **23**:3-4, portrait.

Collection

(1) One Ts: *Entomological Researches in the Marquesas Islands* [Pacific Ocean] by E.P. Mumford and A.M. Adamson. Includes maps and photographs of the Marquesas Islands, c. 1928-1930.

241. MURRAY, Sir John (1841-1914)

Born in Ontario, Canada, Murray was educated in Scotland from the age of 16. He spent twelve years at the University of Edinburgh, but never took his degree.

He was interested in oceanography, and funded several Scottish marine laboratories and undertook several bathymetrical surveys himself. He is most widely known as the naturalist on the *Challenger* expedition, 1873-1976. In 1881, Murray was appointed chief editor of the *Challenger Reports*. He received many distinguished awards including a KCB in 1898 and was elected Fellow of the Royal Society.

Biographical Reference: 1914. W.E.H. *Memoirs and Proceedings of the Manchester Literary and Philosophical Society* **58**:xi-xiii.

Collection

(1) Twelve Ms leaves: *A list of insects including Myriapoda sent to the British Museum. 18.7.1882.*

Insect Room Lists: Volume 2, items 65-76.

242. MYERS, J. G.

Employed by the Imperial Bureau of Entomology, Myers developed a special interest in cicadas. In 1929 he published *Insect Singers : A Natural History of the Cicadas.*

Collection

(1) One annotated Ts: *Insect Singers : A Natural History of the Cicadas.*

(2) One loose-leaf Ms: *The History of Biological Control.*

(3) Miscellaneous notes.

243. NEVINSON, Basil George (1852-1909)

Born in Leicester, he was educated at Shrewsbury School and Exeter College, Oxford. Nevinson was a barrister by profession, and in addition a very keen entomologist, with a particular interest in the Lamellicornia, especially the genus *Phanaeus*. His major publication, *Revised Synonymic List of Species in the Genera* Phanaeus, *Macleay, and* Oxysternon, *Castelnau* was published in 1892. Nevinson was a Fellow of the Entomological Society of London and a member of the Société Entomologique de France.

Biographical Reference: 1910. Fowler, W.W. *The Entomologist's Monthly Magazine* **46**:93-94.

Collection

(1) Ms notebook: *Phanaeus,* c. 1877-1885.

244. NICÉVILLE, Charles Lionel Augustus de (1852-1901)

Born at Bristol, he went to India in 1876 and travelled to many parts of the country. He amassed a fine collection of Indian butterflies and published many papers, mainly in the *Journal of the Asiatic Society of Bengal* and the *Journal of the Bombay Natural History Society*. Nicéville's most important work was *The Butterflies of India, Burmah and Ceylon*, published in 3 volumes (1882-1890). The first volume was co-authored with G.F.L. Marshall, volumes 2 and 3 were by Nicéville alone. Late in his career, Nicéville was appointed Government Entomologist at the Indian Museum. He died from fever in Calcutta.

Biographical Reference: 1903. Anon. *Journal of the Bombay Natural History Society* **14** :140-141.

Collection

(1) Correspondence collection: Approximately 50 letters from C.L.A. de Niceville to Thomas Reid Davys Bell (1863-1948), dated 1880 to 1901.
(2) 22 plates of original watercolour drawings by G.C. Chuckrabutty and B.L. Dos. These were used to illustrate various papers by de Nicéville and B.M. Watson published in the *Journal of the Bombay Natural History Society*. Includes illustrations of Western Himalayan, Indo-Malayan and Indian Butterflies, c. 1890.
(3) Six plates of original watercolour drawings and seven proof plates of butterflies of India, 1882-1890.
(4) Approximately 100 original drawings of Indian butterflies by G.C. Chuckrabutty, B.L. Dos, G.C. Dudgeon, K.C. Hooper and P. Preisss, for Nicéville publications.
(5) Approximately 100 original drawings of Indian butterflies, published in various papers by Nicéville. Includes art work by Nicéville, W. Doherty, H.J. Elwes and J. Wood-Mason.

245. NICHOLL, Mrs.

Collection

(1) Two MS leaves: *Localities of insects in N.W. America*, B.M.1905-69.

Insect Room Lists: Volume 1, item 12A.

246. NORGUET, Anatole de Madré (d. 1898)

Collection

(1) One bound Ms : *Révision du Catalogue des Coléoptères du Nord*. This was prepared as a 2nd edition and 2nd supplement to the *Catalogue des Coléoptères du Département du Nord* (1863).

247. OBERTHÜR, Charles (1845-1924)

Charles and his brother, René, both developed an early interest in entomology; the former in Lepidoptera and the latter in Coleoptera. They formed extensive collections of these groups. Charles' publications included *Études d'entomologie* (1876-1902) and *Études de Lépidoptèrologie comparée* (1904-1924). Oberthür was a member of the Société Entomologique de la France and a Fellow of the Entomological Society of London. The Natural History Museum, London purchased most of the Oberthür collection, which contained over 750,000 specimens of Lepidoptera, in 1927. B.M.1927-3.

Biographical Reference: 1924. Riley, N.D.*The Entomologist* **57**:191-192.

Collection

(1) One loose-leaf Ms: *List of specimens disposed of by C. Höfer prior to purchase of remainder by the British Museum (Natural History)*, c. 1927.

248. OBERTHÜR, René (1852-1944)

René Oberthür was a Fellow of the Royal Entomological Society and in 1931 was elected an honorary Fellow. He was an eminent coleopterist and amassed a fine beetle collection.

Biographical Reference: 1945. Collart, A. *Bulletin et Annales de la Société Entomologique de Belgique* **81**:185. 1945. Cockayne, E.A. *Proceedings of the Royal Entomological Society* of *London* (C) **9**:47.

Collection

(1) Six letters from R. Oberthür to Charles Owen Waterhouse (1843-1917), dated 1903.

249. ORMISTON, Walter

Ormiston spent over 30 years in Ceylon (now Sri Lanka). During this time he collected Lepidoptera, mostly in Haldummulla in the province of Uva. Later he spent two years in England and studied the collections at the Natural History Museum, London. His publications include *The Butterflies of Ceylon*, 1924. He was a Fellow of the Entomological Society of London.

Collection

(1) Correspondence collection: approximately 75 letters, including letters from:

Riley, Norman Denbigh (1890-1979)	1920-1926
Toxopeus, Lambertus Johannes (1894-1951)	1924-1925

250. OSTEN-SACKEN, Carl Robert (1828-1906)

Born in St. Petersburg, he developed an early interest in Diptera and especially the Tipulidae. Osten-Sacken's diplomatic career took him to Washington, where he lived from 1856 to 1877. During this time he collected Diptera and in 1862 he published Catalogue of the described Diptera of North America, Vol. 3 of *Smithsonian Miscellaneous Collections*. A later edition of this work appeared in 1878, *Smithsonian Miscellaneous Collections* no. 270. Osten-Sacken published many other papers and has been called the Father of American Dipterology. His work on the Tipulidae established the outlines of the modern classification of the family. He also studied insect galls and carried out important work on the Tabanidae. He corresponded with Hermann Loew, supplying him with specimens, and translated and published Loew's work in the Monographs of the Diptera of North America (1862-1873), *Smithsonian Miscellaneous Collections*, Nos 6, 171, 219 and 256. He returned to Europe in 1877 and died at Heidelberg.

Biographical References: 1904. Osten-Sacken, C.R. *Record of my Life Work in Entomology*. vii, 240pp. Cambridge, Mass.; 1906. Verrall, G.H. *The Entomologist's Monthly Magazine* **42**:234-235.

Collection

(1) Three Ms leaves: [Notes on Osten-Sacken's Report on the Diptera of the Arctic Expedition], c. 1876.

251. PADMAN, Isaac

Collection

(1) One Ms notebook: *An Essay towards Establishing the Natural History of the English Ant compil'd partly from quotation, chiefly from observation* (second volume only), 1796.

252. PAGANETTI-HUMMLER, Gustav (1871-1949)

In the early part of the twentieth century Hummler was a travelling collector, particularly in Southern Europe. He co-operated with Hans Strichel to edit *Neue Beitrage zur systematischen Insektenkunde* (1916-1932).

Biographical Reference: 1950. Heikertinger, F. *Koleopterologische Rundschau* **31**:151.

Collection

(1) Correspondence collection: Four letters to C.O. Waterhouse, 1902-3, concerning the Paganetti-Hummler collection, together with notes.

253. PAGDEN, Hugh Trelawny (1905-1977)

Born in West Byfleet, Surrey, Pagden studied at University College, London from 1924 to 1927. He was appointed to the Colonial Service and was posted to the Federation of Malay States as Assistant Entomologist, and then to the British Solomon Islands as Senior Entomologist from 1933 to 1935. During World War II he joined the Malayan Volunteer Force and then the RAF. He spent his retirement in Malaya. Pagden was a Fellow of the Royal Photographic Society.

Collection

(1) One box of entomological Mss and miscellaneous notes.
(2) Approximately 500 b/w photographs of Malayan Hymenoptera.
(3) Correspondence collection: approximately 1,000 letters to and from:

Batcheler, D.M.	1954-1960
Fennah, Ronald Gordon (1910-1987)	1964
Hall, Wilfred John (1893-1965)	1954-1960
Oldroyd, Harold (1913-1978)	1952-1955
Parker, A.M.	1970-1975
Pearson, Eric Omar (1906-1968)	1964

(4) Card index: Localities in Malaya and index to Aculeate Hymenoptera Collection.
(5) Photographic collection: includes approximately 1800 projection slides, mainly of Aculeate Hymenoptera; many taken in Penang, Malaya, 1957-1976.
(6) Four Ms notebooks: includes details of Aculeate Hymenoptera collected from Malaya and the Solomon Islands, 1928-1969.

254. PALMER, Mervyn Grove (1879-1955)

During the early part of his career, Palmer went on a number of collecting trips to Central and South America. Lionel Walter Rothschild employed him as a collector. Most of the many hundreds of bird specimens collected by him were subsequently sold by Rothschild to the American Museum of Natural History. Many of the insects Palmer collected came either directly to The Natural History Museum, London, or via the Rothschild collection. Palmer was a Fellow of the Royal Entomological Society of London. He established and was curator of the Ilfracombe Museum in North Devon, which opened to the public in 1932.

Biographical Reference: 1956. Hall, W.J. *Proceedings of the Royal Entomological Society of London* (C) **21**:65.

Collection

(1) Six Ms collecting diaries: Contain details about material collected by Palmer in Central and South America, 1906-1909, as follows:
- (i) *Notebook No. 3* [sic]: [Nicaragua], 1906.
- (ii) *Notebook No. 3: Ecuador,* 1910.

(iii) *Notebook No. 4: Termination of Nicaragua commencement of Colombia,* 7 December 1906 to 11 April 1907.
(iv) *Notebook No. 5: W. Colombia,* 14 April 1907 to 15 October 1907.
(v) *Notebook No. 6*: 28 October 1907 to 24 March 1908.
(vi) *Notebook No. 9: Colombia* 12 September to 7 December 1909.

(2) Correspondence collection: Three letters between Palmer's niece, Barbara Forth and Paul E.S. Whalley of The Natural History Museum, London concerning the purchase of the Palmer diaries, 1967.

255. PARAVICINI, Ludwig (c.1868-1937)

Paravicini lived in Basel, Switzerland. In 1937 The Natural History Museum, London purchased the Paravicini collection, consisting of over 67,000 specimens of Heterocera, including the Oberthür collection of Pyralidae and numerous types. B.M.1937-383.

Collection

(1) One Ms notebook, 1903-1923.
(2) Five letters, 1928-1931, from Johann Müller-Rutz.
(3) One loose-leaf Ms: *Numbers of Pyrales in Coll. Paravicini. 1936 by G.A. Bisset.*
(4) Loose-leaf index: *Index to Paravicini collection Microlepidoptera, (not complete).*
(5) Miscellaneous notes.

256. PASCOE, Frances Polkinghorne (1813-1893)

After a career as a naval surgeon, Pascoe turned his attention to entomology. From 1864 to 1865 he was President of the Entomological Society of London. He travelled through Europe, North Africa and the Amazon, collecting Coleoptera. Pascoe published works on Longicornia, Colydidae, Curculionidae and other families. His collection of over 42,000 specimens, including many types was sold to The Natural History Museum, London. B.M.1893-60.

Biographical Reference: 1893. McLachlan, R. *The Entomologist's Monthly Magazine* **29**:194-196.

Collection

(1) Agreement between Pascoe and the Trustees of The British Museum for sale and purchase of a collection of Coleoptera and cabinets, 1892.

257. PEARCE, Edmund James (1903-1982)

Pearce developed a keen interest in natural history while at school in Sherborne. After graduating from Cambridge he was ordained into the Church of England. His interest in entomology concentrated on the Coleoptera, especially the families Pselapidae and Haliplidae on which he was an international authority. He published extensively on British Coleoptera; his collections went to the National Museum of Wales, Cardiff and The Natural History Museum, London.

Biographical Reference: 1983. MacKechnie-Jarvis, C. *The Entomologist's Monthly Magazine* **119**:171-176.

Collection

(1) Six British collecting log books, 1920-1952 and 1969-73.
(2) Three boxes Ms: *Catalogue of the Pearce Collection Haliplidae*. c. 1970.
(3) One Ms notebook: *Material deposited or sent to the British Museum 1968-* .
(4) Miscellaneous photographs and slides.
(5) Loose-leaf Ms: Earliest collecting records, c. 1920.

258. PEARMAN, John Victor (1887-1970)

Pearman was born in Linslade, Buckinghamshire. Between 1923 and 1960 he published 40 papers on the Psocoptera. These included a preliminary sketch of a classification of Psocids. He was a member of the Bristol Naturalists Society and a Fellow of the Royal Entomological Society of London.

Biographical Reference: 1971. Clay, T. and Smithers, C.N. *The Entomologist's Monthly Magazine* **107**:65-66.

Collection

(1) Seven boxes of Ms notes and drawings on Psocoptera.

Box 1:

- (i) Ms notes: on the preparation of a Psocoptera Handbook (unpublished).
- (ii) Notes on venation.
- (iii) Nomenclatural notes.
- (iv) Descriptions and figures of New Hebrides (Pacific Ocean) New Guinea and Papuan species.
- (v) Notes on Wegener Theory – distribution of some Psocids.
- (vi) Notebook: *Psocoptera, systematic and nomenclatural.*
- (vii) One file: *Historical notes on deathwatch beetles, Taxonomy Groundwork.*
- (viii) One file: Notes and drawings on Pseudocaecilius.
- (ix) One notebook: *Pachytroctid analysis, and notes on* Philotarsus *complex.*
- (x) One notebook: *Myopsocus* descriptions and drawings.
- (xi) Notes on biometry, varieties, morphology and systematics.
- (xii) One notebook : List of genera and species, when described and by whom, with distribution notes.
- (xiii) Miscellaneous notes and drawings, mostly descriptions.

Box 2:

- (i) *Notes on some aberrant genera of Psocina.*
- (ii) *Draft catalogue.*
- (iii) One packet: Notes on *Liposcelis,* including notes on a tentative classification.
- (iv) One file: Notes on microscopy and techniques applicable to Psocids.
- (v) One file: Extracts from papers published in various journals by other authors.

Box 3:

- (i) One packet: Notes and figures on *Troctes* var.

Box 4:

- (i) *Notes on the relationship of the genera of the Elipsocidae.*
- (ii) *Island Psocids* – data regarding distribution of genera.
- (iii) *Liposceliss*: Descriptions, nomenclature, morphology.
- (iv) One notebook: [Distribution of *Caecilius*.]
- (v) Book of drawings: Pachytroctidae.
- (vi) References to old literature.
- (vii) List of Stephens types in British Museum.
- (viii) Extracts from papers.
- (ix) Historical notes on some insects, including deathwatch beetle.

Box 5:

- (i) Ms notes on the Pacific Entomological Survey, including: ecological notes, collections and preservation, preliminary descriptions.
- (ii) Notes on Solomon Islands materials and Australian Psocids.

Box 6:

- (i) Working notes on Psocoptera.
- (ii) Family tables, notes on booklice.
- (iii) One notebook: Colour drawings of wings.
- (iv) Miscellaneous notes and drawings.

Box 7:

- (i) One notebook: *Classification of genera of Psocoptera*.
- (ii) One notebook: [Illustrations with legends of Psocids eggs and young emerging from eggs.]
- (iii) Ms: Identities of some species of Psocoptera.
- (iv) Six notebooks: Extracts of papers on Psocids.
- (v) Miscellaneous notes and drawings.

(2) Two boxes (nos 8 and 9) of correspondence : Letters, 1925-1956, to and from 123 correspondents, including:

Balfour-Browne, William Alexander Francis (1874-1967)	1934
Baweja, K.D. (1896-1962)	1936-1937
Bequaert, Joseph Charles (1886-1982)	1949
Carpenter, Geoffrey Douglas Hale (1882-1953)	1942-1943
Cheesman, Lucy Evelyn (1881-1969)	1930-1937
Chopard, Lucien (1855-1971)	1955

Clay, Theresa	1936-1957
Cowley, John (1909-1967)	1947-1948
Crampton, Guy Chester (1881-1951)	1929
Eastop, Victor Frank	1953
Eltringham, Harry (1873-1941)	1933
Enderlein, Günter (1872-1968)	1930
Evans, John William (1906-1990)	1945
Finlayson, Lawrence H (1903-1965)	1944-1945
Fox-Wilson, George (1896-1951)	1932
Gambles, Robert Moylan (1910-1990)	1931-1946
Green, Edward Ernest (1861-1949)	1932-1937
Haines, Frederick Haslewood (1864-1946)	1934
Hall, Wilfred John (1893-1965)	1949-1958
Hemming, Arthur Francis (1893-1964)	1950-1951
Hindle, Edward (1886-1973)	1948
Hinton, Howard Everest (1912-1977)	1943-1944
Hobby, Bertram Maurice (1905-1983)	1945-1955
Imms, Augustus Daniel (1880-1949)	1929-1936
Karney, Heinrich Hugo (1886-1939)	1931-1935
Kearns, Howard George Henry (1902-1986)	1937-1938
Keys, James H (1855-1941)	1928
Killington, Frederick James (1894-1956)	1934
Kimmins, Douglas Eric (1905-1985)	1929-1934
King, C.B. Redman	1935-1940
Kirkpatrick, Thomas Winfred (d. 1971)	1932
Lloyd, Robert Wyllie (1868-1958)	1932
Longfield, Cynthia (1896-1991)	1948
Macan, Thomas Townley (1910-1985)	1936
MacBride, Ernest Williams (1866-1940)	1833-1935
McKenny Hughes, Alfred Weston (1895-1870)	1934
Massee, Arthur Morel (1899-1967)	–
Munro, J. W. (1888-1968)	1929
Musgrave, Anthony J. (1895-1959)	1937
Neave, Sheffield Airey (1879-1961)	1931-1945
Peach, Alec Hamilton (1885-1960)	1956

Pearson, Eric Omar (1906-1968)	1957
Perkins, Robert Cyril Layton (1866-1955)	1919
Potter, Charles (1907-1959)	1938
Poulton, Edward Bagnell (1856-1943)	1928
Priesner, Hermann (1891-1974)	1936
Richards, Owain Westmacott (1901-1984)	1928
Riley, Norman Denbigh (1890-1979)	1955
Taylor, Thomas Hugh Colebrook (1901-1972)	1947-1952
Vesey-Fitzgerald, Leslie Desmond Foster (1910-1974)	1936
Weber, Hermann (1899-1956)	1955
Williams, Carrington Bonso (1889-1981)	1932-1944
Zacher, Fredrich	1939
Zimmerman, Elwood C.	1939-1947
Zumpt, F. (1908-1985)	1956

(3) One box (no. 10) of drawings: Pen-and-ink drawings of Psocoptera.

259. PEILE, Harry Diamont (1872-1959)

Peile served 24 years in the Indian Army, including a period as medical officer to the 52nd Sikhs on the N.W. Frontier of India, and achieved the rank of Lieutenant-Colonel. He was a keen lepidopterist and in 1937 published his *Guide to Collecting Butterflies of India*. During World War I he served in Iraq, and from his observations made important contributions to the knowledge of butterflies in Mesopotamia and N.W. Persia. He presented his collection to The Natural History Museum, London.

Biographical Reference: 1959. Hutchinson, C.H. *The Entomologist* **92**:244.

Collection

(1) One Ms notebook: Entomological notes and observations, photographs of places and historical events. It includes:

(i) Collecting diary from Peshawar and Mussoorie, India, 1911-1917.

(ii) Notes on a journey from Baghdad to Babylon.

(iii) Notes on Menton, S. France 1928-1930.

(iv) Drawings and nature prints.

260. PERKINS, John Frederick (1910-1983)

Perkins studied entomology at Imperial College, London from 1928 to 1932. In 1933 he joined the staff of The Natural History Museum, London. He spent his career working on parasitic Hymenoptera, specializing in the Ichneumonidae. When Perkins retired in 1973 he was Deputy Keeper of the Department of Entomology. He was a Fellow of the Royal Entomological Society.

Biographical Reference: 1983. Quinlan, J. and Huddleston, T. *Antenna* **7**:120-121.

Collection

(1) Manuscript and drawings relating to the *Handbooks for the Identification of British Insects*, Vol. VI. Part 3(a) Hymenoptera. Bethyloidea. (1976).

261. PERKINS, Robert Cyril Layton (1866-1955)

Perkins read classics at Oxford, but under the influence of E.B. Poulton changed to science. In 1891 he was chosen by the Members of the Sandwich Island Committee (Royal Society and the British Association for the Advancement of Science), to investigate the fauna of the Hawaiian Islands. For 10 years, 1892-1902, he collected specimens, which resulted in his contributions to *Fauna Hawaiiensis* (1899-1913). When the collection was divided up The Natural History Museum, London received the type specimens; the remaining specimens went to the respective specialists and the Bernice P. Bishop Museum, Honolulu. In later years his interests turned to the biological control of imported insect pests and introduced weeds, on which he worked with Frederick Muir, O.H. Swezey and Albert Koebele. In 1912 he retired and continued to work on his collections. His work on British insects concentrated on Aculeate Hymenoptera and sawflies. He published over 200 papers in the *The Entomologist's Monthly Magazine.*

Biographical References: 1955. Scott, H. and Benson, R.B. *The Entomologist's*

Monthly Magazine **91**:289-291; 1986. Manning, A. *Bishop Museum Occasional Papers* **26**:1-46.

Collection

(1) Collecting notes from various Hawaiian localities, 1892-1936.
(2) Diary of collecting on Kanai, 1895.
(3) Autobiographical notes.
(4) Notes on Hymenoptera collected in the Marquesas Islands, Pacific Ocean.
(5) Reports of the Commissioner, 1898-1902. List of injurious and benefical insects of the Hawaiian Islands.
(6) Miscellaneous notes and drawings.
(7) One Ms notebook: *Gloucester Lepidoptera, Wotton-under-Edge, 1891.*
(8) Seven examination certificates from Jesus College, Oxford.
(9) Correspondence collection: 300 correspondents, 1885-1948, including the following:

Adamson, Alastair Martin (1901-1945)	1932
Alexander, Charles Paul (1899-1981)	1912
Alfken, Johann Diedrich (1862-1945)	1913-1922
Arrow, Gilbert John (1873-1948)	1914
Ashby, Edward Bernard	1925-1935
Ashmead, William Harris (1855-1908)	1901-1905
Atkinson, Harold W.	1936-1937
Audcent, Henri L. F. (1875-1951)	1931
Auderson, Esther C.	1932
Austen, Ernest Edward (1867-1938)	1906-1931
Bagnell, Richard Siddoway (1889-1962)	1908
Balfour-Browne, John	1925-1931
Balfour-Browne, William Alexander Francis (1874-1967)	1930
Banks, Nathan (1868-1953)	1908-1909
Bateson, William (1861-1926)	1909-1915
Bedwell, Ernest Charles (1875-1945)	1929
Benson, Robert Bernard (1904-1967)	1923-1935
Bethune-Baker, George Thomas (1857-1944)	1915-1920
Bezzi, Mario (1868-1927)	1880-1927

Bignell, George Carter (1826-1910)	1891
Billups, Thomas Richard (1841-1919)	1888
Blair, Kenneth Gloyne (1882-1952)	1919-1931
Blüthgen, P. (1880-1967)	1926
Boving, Adam (1870-1957)	1909
Bridgeman, John Brooks (1836-1899)	1888
Britten, Harry (1870-1954)	1921-1946
Brues, Charles Thomas (1879-1955)	1905-1909
Bullock, Edwin (1879-1965)	1928
Burgess, Albert Franklin (1873-1953)	1909
Burr, Malcolm (1878-1954)	1897-1898
Burtt, Bernard O.	1922-1923
Butterfield, Rosse (1875-1935)	1919-1921
Buxton, Patrick Alfred (1892-1955)	1927
Calvert, Philip Powell (1871-1961)	1911-1913
Carpenter, Geoffrey Douglas Hale (1882-1953)	1938-1942
Carr, John Wesley (1862-1939)	1924
Champion, George Charles (1851-1927)	1911-1918
Chapman, Christopher W. H.	1943
Chapman, Lucille	1944
Chawner, Ethel Frances (1866-1953)	1928-1934
Cheesman, Lucy Evelyn (1881-1969)	1926-1937
China, William Edward (1895-1979)	1948
Cockerell, Theodore Dru Alison (1866-1948)	1910-1924
Collenette, Cyril Leslie (1888-1959)	1928
Cowley, John (1909-1967)	1946
Craw, Alexander (1850-1908)	1903
Crawford, James Chamberlin (1880-1950)	1909
De Worms, Charles Maurice ,Baron (1903-1979)	1936
Donisthorpe, Horace St John Kelly (1870-1951)	1916-1948
Donitz, W.	1909
Durrant, John Hartley (1863-1928)	1891-1909
Edwards, Frederick Wallace (1888-1940)	1915
Ehrhorn, Edward Macfarlane (1862-1941)	1914
Elliott, Ernest A. (d. 1936)	1915

Ellis, Henry Willoughby (1869-1943)	1925
Eltringham, Harry (1873-1941)	1920
Enderlein, Günther (1872-1968)	1913
Enock, Frederick (1845-1916)	1914
Fassnidge, William (1888-1949)	1927
Forel, Auguste Henri (1848-1931)	1913
Fraser, Frederic Charles (1880-1963)	1948
Friend, Roger B.	1930
Froggatt, Walter Wilson (1858-1937)	1906
Fullaway, David Timmins (1880-1964)	1913-1917
Gahan, Charles Joseph (1862-1939)	1916-1921
Gardner, Willoughby (1860-1953)	1919-1935
Giffard, Walter M. (1856-1929)	1904-1925
Green, Edward Ernest (1861-1949)	1928-1930
Griffith, Arthur Foster (d. 1933)	1932
Haines, Frederick Haslewood (1864-1946)	1914-1921
Hallett, Howard Mountjoy (1878-1958)	1914-1948
Haviland, Maud Doria (1891-1941)	1927
Hofender, Karl (1878-1951)	1908
Holland, William Jacob (1848-1932)	1905
Horn, Walter (1871-1939)	1919
Horvath, Gezza de (1847-1937)	1916
Howard, Leland Ossian (1857-1950)	1916
Imms, Augustus Daniel (1880-1949)	1907
Jacobson, Edward (1870-1944)	1907-1909
Jermyn, Turenne (1862-1924)	1923-1924
Johnson, William Frederick (1852-1934)	1920-1027
Jones, Hugh Parry (1893-1937)	1925-1931
Joy, Ernest Edward Cooper (1869-1940)	1930
Joy, Norman Humbert (1874-1953)	undated
Kemp, Stanley (d. 1945)	1943
Kennedy, Clarence Hamilton (1870-1952)	1925
Kertész, Kálmán (1867-1922)	1905-1915
Keys, James H. (1855-1941)	1922
Kieffer, Jean Jacques (1856-1925)	1905-1913

Killington, Frederick James (1894-1956)	1931
King, James Joseph Francis Xavier (1855-1933)	1914-1917
Kirkaldy, George Willis (1873-1910)	1904-1910
Koebele, Albert (1852-1924)	1897-1913
Konow, Fredrick Wilhelm (1842-1908)	190[5]
Kuwana, Shinkai Inokichi (1872-1933)	1905
Laidlaw, F. F. (1876-1963)	1925
Lallemand, Victor (1880-1965)	1913
Lamb, Charles George (1861-1941)	1926
Latter, Oswald H. (1864-1948)	1919-1921
Lea, Arthur Mills (1868-1932)	1906
Lebeden, Alexander G. (1874-1936)	1929
Marlatt, Charles Lester (1863-1954)	1906
Marshall, Guy Anstruther Knox (1871-1959)	1912-1932
Mason, Philip Brookes (1842-1903)	1888
May, Jan (d. 1959)	1937
Meade-Waldo, Geoffrey (1884-1916)	1909-1916
Meyrick, Edward (1854-1938)	1895-1908
Miyake, Tsunekata (1880-1921)	1915
Montandon, Arnold Lucien (d. 1922)	1908
Morice, Francis David (1849-1926)	1890-1926
Morley, Claude (1874-1951)	1947
Mortimer, Charles Henry (d. 1932)	1924
Morton, Kenneth J. (1858-1940)	1920-1930
Mosely, Martin Ephraim (1867-1948)	1930-1933
Muir, Frederick Arthur Godrey (1872-1931)	1904-1931
Mumford, Edward Philpott	1931-1936
Murray, James (1872-1942)	1929-1938
Muttkowski, Richard Anthony (1887-1943)	1909-1910
Nevinson, Edward Bonney (1858-1927)	1919
Nixon, Gilbert Edward James (1905-1987)	1925-1938
Nurse, Charles George (1862-1933)	1914-1921
Pagden, Hugh Trelawny (1905-1977)	1925-1930
Poulton, Edward Bagnell (1856-1943)	1897-1937
Pratt, Frederick C. (1869-1911)	1907

Rait-Smith, William (1875-1958)	1934
Richards, Owain Westmacott (1901-1984)	1926-1935
Riel, Philibert (1862-1943)	1923
Riley, Norman Denbigh (1890-1979)	1929-1937
Robbins, John Cuthbert (1906-1932)	1928-1931
Roberts, Arthur William Rymer (1879-1955)	1917
Rothschild, Nathaniel Charles (1877-1923)	1919-1920
Routledge, George Bell (1864-1934)	1917-1918
Sampson, Frank Winn (1853-1926)	1920
Saunders, Charles James (1868-1941)	undated
Saunders, Edward (1848-1910)	1889-1905
Saunt, John William (1881-1958)	1933-1940
Schouteden, Henri (1881-1972)	1909-1912
Schulthess, Anton (1855-1941)	1913
Scott, Hugh (1885-1960)	1909-1943
Shestakov, Andrei Valentinovich (1890-1933)	undated
Silvestri, Filippo (1873-1949)	1903-1918
Spooner, Malcolm (1907-1989)	1930-1947
Stelfox, Arthur Wilson (1883-1972)	1922-1935
Stephenson, T. A.	1916
Stoeckhert, E.	1925
Swain, Humphrey Drummond (1902-1959)	1927
Swale, Harold (1853-1919)	undated
Swenk, Myron Harmon (1883-1941)	1908
Swezey, Otto Herman (1869-1959)	1906-1947
Taylor, Thomas Hugh Colebrook (1901-1972)	1924
Terry, Frank Wray (1877-1911)	1904-1908
Thornley, Alfred (1855-1947)	1924-1934
Turner, Henry Jerome (1856-1951)	1928
Turner, Rowland Edwards (1863-1945)	1908-1910
Varley, George Copley (1910-1982)	1932
Viereck, Henry Lorenz (1881-1931)	1908-1925
Wainwright, Colbran Joseph (1867-1949)	1934
Walker, James John (1851-1939)	1917
Walsingham, Thomas de Grey (1843-1919)	1908

Waterhouse, Charles Owen (1843-1917)	1907-1914
Waterhouse, Gustavus Athol (1877-1950)	1906
Waters, Edwin George Ross (1890-1930)	1928
Waterston, James (1879-1930)	1915-1929
Williams, Francis Xavier (1882-1967)	1935
Wood, Theodore (1862-1923)	1892
Wormersley, Herbert (1889-1962)	1924
Zimmerman, Elwood C.	1936-1950

(10) Ms: *Keys to numbers of localities of Mr Perkins' Hawaiian insects, First Expedition 1892-94, and 2nd Expedition 1895-97.*

Insect Room Lists: Volume 1, items 3, 39-59.

262. PETHER, W. G.

Collection

(1) One Ms: *Catalogue of the Pether collection of Lepidoptera.*
(2) One Ts: [Catalogue of collection of Palaearctic Rhopalocera.]

263. PICTET, François Jules (1809-1872)

Pictet was born in Geneva, to a wealthy Swiss family. In 1830 he went to Paris and became one of the first entomologists to specialize in the study of the Ephemeroptera. Much of his work formed the foundation for research on this group. Between 1841 and 1845 he published in a total of 21 parts, a two-volume work: *Histoire naturelle générale et particulière des insectes néuroptères.* In his later years he devoted himself to palaeontology.

Biographical Reference: 1990. Peters, J. *et al.*, in Flannagan, J.F. and Marshall, K.E. *Advances in Ephemeroptera Biology.* London. 552pp. (pp. 532-534).

Collection

(1) Two Ms leaves: [Lists Trichoptera type specimens given to Curtis, probably written by J. Curtis]

264. PIERCE, Frank Nelson (1861-1943)

Pierce was a keen field collector and a member of the Lancashire and Cheshire Entomological Society. His major work started in 1909 when he published *The Genitalia of the Group Noctuidae of the Lepidoptera of the British Islands*. Subsequently he published other volumes covering the genitalia of all British Lepidoptera.

His collection of British Lepidoptera and the drawings for his work, *The Genitalia of the Group Noctuidae of the Lepidoptera of the British Isles*, were bequeathed to the Liverpool Museum and the slide preparations were donated to The Natural History Museum, London.

Biographical Reference: 1943. W.M. *The Entomologist* **76**:175-176.

Collection

(1) 23 watercolour Christmas cards drawn by Pierce, 1904-1941.
(2) One Ms notebook.
(3) Seven letters to Harold Williams Daltry, 1937-1938.

265. PITTIER DI FABREGA, M.H.

Collection

(1) One Ms leaf: [List of Coleoptera presented by M.H. Pittier di Fabrega from Costa Rica], 1896.

Insect Room Lists: Volume 2, item 62.

266. POOLE, Herbert F.

Poole was a keen lepidopterist and contributed a chapter on the Lepidoptera to F. Morey's, *A Guide to the Natural History of the Isle of Wight* (1909).

Collection

(1) One Ms notebook: *Natural History lists and letters relating to the Guide,* 1908.
(2) One Ms notebook: *The Lepidoptera of the Isle of Wight, Times of emergence - note book no.3*, 1910-1929.

267. POPE, Robert Denis (b. 1928)

Educated at Selwyn College, Cambridge where he gained his MA in 1977. He was employed by the Commonwealth Institute of Entomology, 1949-1964, and The Natural History Museum, London until 1986. Pope's main areas of interest have been and continue to be Coccinellidae, Colydiidae, the British Beetle Fauna, nomenclatural problems and the micromorphology of insect waxes. He has published several books and papers.

Collection

(1) Two Ms field notebooks: [Notes on Lopham and Redgrave Fens, Suffolk and Norfolk.]

268. POWER, John Arthur (1810-1886)

Power studied at Clare College, Cambridge and practised medicine for 40 years. While at Cambridge he began to study Coleoptera. Between 1853 and 1880 he collected regularly and found many species of Coleoptera and Hemiptera new to the British lists. His collection of 30,000 British Coleoptera and 6,000 Hemiptera was purchased by The Natural History Museum, London in 1896. B.M.1896-69.

Biographical Reference: 1886. Dunning, J.W. *The Entomologist* **19**:193-200.

Collection

(1) Two Ms notebooks: *Register of Power's* [British] *Collection 1848-1879*.

269. PRATT

Collection

(1) 189 b/w photographs of butterflies from various localities collected by Messrs. Pratt, c. 1910.

270. PROUT, A. Ellen

Collection

(1) One loose-leaf Ms: *Hypenid Genera*, c. 1930.

271. PROUT, Louis Beethoven (1864-1943)

The world authority on the Geometridae, he contributed fascicules on this group for Wytsman's, *Genera Insectorum* (1902-1970) and Seitz's, *Die Gross-Schmetterlinge der Erde* (1906-1954). He presented his collection to The Natural History Museum, London. Prout was President of the City of London Entomological and Natural History Society, 1899-1904, and a Special Life Fellow of the Royal Entomological Society of London.

Biographical Reference: 1944. Cockayne, E.A. *Entomologist's Record and Journal of Variation* **56**:28.

Collection

(1) 25 boxes of Ms notes on the Geometridae.
(2) Six Ms notebooks: *Entomological Journals*, 1889-1921.
(3) Three boxes of miscellaneous entomological notebooks.
(4) Two Ms notebooks: *Reference catalogue of Lepidoptera literature*.
(5) Six Ms notebooks: *Notes on Larentidae*.
(6) 13 Ms notebooks: *Conspectus nominum Geometrarum*.

(7) 699 genitalia drawings together with notes on *Eumelia.*

(8) 12 miscellaneous Ms notebooks and drawings of Lepidoptera genitalia.

(9) Loose-leaf Mss: *Notes on collections (Geometrid, Hill Museum).*

(10) One Ts: [Geometridae].

(11) Three boxes of miscellaneous Mss.

(12) Approximately 1000 letters sent to Prout, 1900-1940, from, among others:

Bayne, Arthur F. (1869-1947)	1903-1910
Blackmore, Ernest Henry (1878-1929)	1919-1922
Burrows, Charles Richard Nelson (1851-1936)	1909-1917
Forbes, William Trowbridge Merrifield (1886-1968)	1928-1938
Giacomelli, Eugenio (d. 1941)	1910-1920
Grossbeck, John Arthur (1883-1914)	1908-1913
Grote, August Radcliffe (1841-1903)	1900-1902
Harrison, John William Heslop (1881-1967)	1909-1916
Janse, Antonie Johannes Theodorus (1877-1970)	1912-1934
Kirkaldy, George Willis (1873-1910)	1904-1910
Lhomme, Léon (1867-1949)	1926-1929
Littler, Frank Mervyn (1880-1922)	1904
Littlewood, Frank (1882-1949)	1911-1937
McDunnough, James Halliday (1877-1962)	1911-1940
Paoli, Guido (1881-1947)	1927
Pierce, Frank Nelson (1861-1943)	1896-1923
Puengeller, Rudolf (1857-1927)	1907-1912
Quail, Ambrose (1872-1905)	1901-1904
Querci, Orazio (1875-1967)	1920
Rait-Smith, William (1875-1958)	1935
Robertson, R. Bowen (1860-1919)	1912
Schaus, William (1858-1942)	1920-1936
Seitz, Adalbert (1860-1938)	1935-1938
Sterneck, Jakob Doublebsky von (1868-1941)	1934-1939
Taylor, George William (1851-1912)	1904-1911
Warren, William (1839-1914)	1907-1912
Wehrli, Eugène (1871-1958)	1934-1936

272. QUERCI, Orazio (1875-1967)

Born in Rome, Querci developed an early interest in butterflies. He married Clorinda di Nino who shared his entomological interests. He collected for The Natural History Museum, London, Baron Rothschild and Charles Oberthür. In 1927-1928 he carried out a survey of the Lepidoptera of Cuba, on his return to Europe he spent time in Lisbon, Madrid and Tangier.

Biographical Reference: 1968. S.N.A. J[acob]. *The Entomologist's Record and Journal of Variation* **80**:87-88.

Collection

(1) Ts: *Clorinda Querci's beginning and end*. Typescript of an obituary published in *The Entomologist's Record and Journal of Variation* 1959 **71**: 215-217 (1959).
(2) Four family photographs.

273. RAYNOR, Gilbert Henry (1854-1929)

Born in Totternhoe, Bedfordshire, Raynor became a keen lepidopterist. Between 1876 and 1880 he travelled through New South Wales, Victoria and Queensland, Australia. On his return his collection was divided between The Natural History Museum, London and Lord Walsingham. The Walsingham collection came to The Natural History Museum, London as part of the Walsingham bequest. Raynor collaborated with Tutt, contributing to the *Entomologist's Record* and Tutt's volumes on British Lepidoptera.

Biographical Reference: 1929. Riley, N.D. *The Entomologist* **62**:239-240.

Collection

(1) One Ms notebook: *Catalogue of Australian Lepidoptera 1887-8.*

274. REID, W.

Reid lived in Sheffield and developed a keen interest in Lepidoptera. His collection of approximately 10,000 Lepidoptera specimens from the British Isles was purchased by The Natural History Museum, London in 1977. B.M.

1977-487.

Collection

(1) Two Ms notebooks: [Notes on collecting British insects], 1951-1955.

275. RICHARDS, Owain Westmacott (1901-1984)

Educated at Hereford Cathedral School and Oxford University, in 1927 he joined Imperial College, London, to research insect pests of stored products and remained at the College until his retirement in 1967. He was Head of the Department of Zoology and Applied Entomology, and Director of the Field Station at Silwood Park. His entomological interests included the taxonomy of groups of Diptera, including the Sphaeroceridae, and the Aculeate Hymenoptera. Richards published over 180 papers and was the author of several books including, *The Social Insects* (1953), *The Social Wasps of the Americas* (1978) and was the major editor of the revision of *Imms' General Textbook of Entomology* in 1957. He was a Fellow of the Royal Entomological Society of London and its President in 1957-1958.

Biographical Reference: 1985. Davies, R.G. and Waloff, N. *Antenna* **9**:60-62.

Collection

(1) Nine boxes of miscellaneous notes and drawings relating to Aculeate Hymenoptera, c. 1970.
(2) Author's annotated copy of *The Social Wasps of the Americas* (1978).
(3) 29 Ms notebooks: Concerning social wasps, c. 1950s.
(4) Six Ms notebooks: Australian *Ropalidia* and *Polistes*, c. 1950s.
(5) Three Ms notebooks: *Belonogaster*, c. 1950.
(6) Three Ms notebooks: Mainly notes on British bees and wasps.
(7) 11 Ms notebooks: [Notes on *Bombus* and *Psithyrus*], c. 1960s.
(8) Three Ms notebooks: Miscellaneous notes on Diptera and Wasps.
(9) Two Ms notebooks:
    (i) *Arabian Masaridae*, c. 1960.
    (ii) *Records of Matto Grosso Bees*, c. 1950.

276. RILEY, Norman Denbigh (1890-1979)

Born in Tooting, London, Riley developed an early interest in natural history. He studied entomology under Professor Maxwell–Lefroy at the Imperial College of Science and in 1911 joined the staff of The Natural History Museum, London. During World War I Riley served in the Army Service Corps and the Queen's Regiment. On his return to the Museum he continued his work on the butterflies and in particular the Lycaenidae. In 1932 he was appointed Keeper of the Department of Entomology and retained this post for 22 years. Riley was editor of *The Entomologist*. His many publications included: *The Department of Entomology of the British Museum (Natural History) 1904-1964. A Brief Historical Sketch* (1964) – printed for the delegates of the International Congress of Entomology, 1964; *A Field Guide to the Butterflies of the West Indies* (1975) and, with L.G. Higgins, *A Field Guide to the Butterflies of Britain and Europe* (1970). He was a Fellow of the Royal Entomological Society, the Zoological Society and the Institute of Biology. He was also a member of the British Entomological and Natural History Society and the Lepidopterists' Society. For many years Riley served as the Verall Supper member of the prestigious Entomological Club.

Biographical References: 1979. Gilbert, P. *Nature, London* **281**:164; 1979. Howarth, T.G. *et al. Antenna* **3**:130-134.

Collection

(1) Ms material relating to *A Field Guide to the Butterflies of the West Indies* (1975), including:

- (i) Annotated Ts.
- (ii) Notes, general correspondence.
- (iii) Notes on the Hesperiidae.
- (iv) Correspondence with the artist Gordon Riley.
- (v) Correspondence with American entomologists, 1971-1975, mostly with F. Martin Brown, 1971-1972.

(2) Correspondence collection, including letters to and from:

Brown, F. Martin	1966-1972
Sheldon, William George (1859-1943)	1919-1928

Stempffer, W.G. 1938-1974

(3) Correspondence relating to the works, *A Field Guide to the Butterflies of the West Indies* (1975) and *A Field Guide to the Butterflies of Britain and Europe* (1970).
(4) Ts: Material dealing with Cook's second Voyage, by A. Lysaght.
(5) Ts: N.D. Riley's translation of *Belser Butterfly Book* (not published) by Danesch and Dierl.
(6) Miscellaneous correspondence.
(7) Ms notes made by Entomology Staff of Natural History Museum, London of their active service during World War I.
(8) Ms notes: *Early Staff of the Entomology Department and the History and Growth of the Entomology Department B.M.(N.H.).*
(9) Ms and draft: *The Department of Entomology of the British Museum (Natural History) 1904-1964, A Brief Historical Sketch* 1964.
(10) Signature book: *Signatures of friends and colleagues presented to Riley on his retirement from B.M.(N.H.)*, 1955.

277. ROBERT, Charles (1802-1837)

Born into a wealthy family in Liège, Robert developed an early interest in natural history. His particular interest was the study of Diptera, although he also worked on the Hymenoptera and Coleoptera. Robert published a number of papers and built a fine collection of insects from the Liège region of France.

Biographical Reference: 1837. Lacordaire, J.T. *Bulletin de la Société Entomologique de France* **6**:xxxi-xxxiii.

Collection

(1) One Ms notebook: *Catalogue des Coléoptères des environs de Liège.*

278. ROBINSON, Edward W. (1835-1877)

Little is known about Robinson. He was an accomplished zoological artist, illustrating scientific papers in journals such as, *Proceedings of the Zoological*

Society of London.

Biographical Reference: 1986. Root, N.J. and Johnson, B.R. *Proceedings of the Zoological Society of London: An Index to Artists 1848-1900*. 947pp. New York: Garland Publishing.

Collection

(1) 50 drawings of Microlepidoptera attributed to Robinson, c. 1860s.
(2) Pattern plates for an unidentified Lepidoptera work.

279. ROSEN, Baron von

On the death of Baron von Rosen the Trustees of his Munich estate disposed of his important collection of entomological drawings and manuscripts. The collection was purchased for The Natural History Museum, London by the Friends of National Libraries in 1949.

Collection

The collection contains examples of work from a distinguished group of Bavarian and Austrian entomologists from the period 1770-1850. This includes work by Jacob Hübner (1761-1826), Ignaz Schiffermüller (1727-1809) and Gottlieb August Wilhelm Herrich-Schäffer (1799-1874). Other important authors and artists represented include: J.E. Fischer von Röslerstamm (1787-1866), Johann Rudolph Schellenberg (1740-1806), Carl Geyer and Carl Plötz (1813-1886).

List of original files:

File 1:

(i) One hand-coloured engraving by J.R. Schellenberg, apparently sent to I. Schiffermüller.

(ii) 54 sheets of 407 coloured figures of pupae, caterpillars and food plants by Schiffermüller. The drawings are annotated by Herrich-Schäffer and refer to the reproductions in *Systematisches Verzeichniss der Schmetterlinge der Wienergegend*, 1776 by Schiffermüller and J.N.C.M. Denis. These drawings were subsequently placed at the disposal of C.J. Hübner who used them in his *Geschichte Europäischer*

Schmetterlinge [1793]-[1841].

(iii) Ms sheets as follows:

1[a] 4 leaves : *Observations novae Ph. Bombyces.*

1[b] l leaf : *R.P.Piller* [notes by Piller were incorporated in Schiffermüller's notes].

1[c] 2 leaves dated 1771.

1[d] 2 leaves: *in Ladula du Bica. Septmbr [sic] 1775.*

1[e] 1 leaf: *in vitro.*

1[f] 2 leaves: [larval descriptions.]

1[g] 2 leaves: *Vitro 1776.*

2(a) 2 leaves : *A.P.Denis annotata Ph. Noctuae.*

2(b) 2 leaves: *Larvae 1775.*

2(c) 1 leaf: *Verzeichniss* [in Schiffermüller's hand].

(3) [not present]

(4) One uncoloured plate, and 8 leaves of annotated drawings.

(5) Four leaves: *Erläuterungen ueber das systematisches Verzeichniss der Schmetterlinge der Wienergegend aus der Urquelle gezogen* von Jacob Hübner 1797.

6) Annotated double quarto sheet.

File 2: 29 leaves of drawings by Schiffermüller of larvae and pupae, with Latin and German text.

File 3: 107 watercolour copies by Schaedler of Schiffermüller's original drawings, with names added by Hübner.

File 4: 105 pages of Microlepidoptera drawings by F. Roth von Schreckenstein (1753-1808), annotated by Hübner.

File 5: 71 sheets of mounted drawings for Hübner's *Beitrage zur Geschichte der Schmetterlinge* (1786-1790) and *Sammlung europäischer Schmetterlinge* [1783-] 1796-1830[-1841].

File 6: Hübner's Ms notes, with drafts and proofs *ex libris* Herrich-Schäffer.

File 7: 55 sheets of original drawings by J.E. Fischer von Röslerstamm, A. Harzer and J. Mann. Includes some handcoloured, unpublished plates for J.E. Fischer von Röslerstamm's *Abbildungen zur Berichtigung und Ergänzung der Schmetterlingskunde* (1834-[1843]).

File 8: Original wrappers and some plates from J.E. Fischer von Röslerstamm's *Abbildungen zur Berichtigung und Ergänzung der Schmetterlingskunde*

(1834-[1843]).

File 9: Original paintings by Geyer for the continuation of Hübner's *Sammlung exotischer Schmetterlinge* (1806-[1841]). Includes pencilled legends by Geyer and Herrich-Schäffer, and ink notes by [Boisduval].

File 10: Original paintings by Geyer.

File 11: Geyer's drawings for the plates of *Sammlung Exotischer Schmetterlinge* vols. 2-3.

File 12: Geyer's drawings for the plates of vols. *Sammlung Exotischer Schmetterlinge.*

File 13: Original paintings of larvae by Geyer, with copies after Rambur.

File 14: Collection of correspondence includes:

Three letters from Geyer to Herrich-Schäffer, c. 1865-1871.

One letter from Ernst August Helmuth von Kiesenwetter (1820-1880), 1877.

Two letters from Heinrich Gottlieb Ludwig Reichenbach (1793-1879), 1860, 1866.

One letter from Johann Heinrich Christian Friedrich Sturm (1805-1862), 1853.

Two letters from Eugenius Johann Christoph Esper (1742-1810), 1796, 1805.

File 15: Drawings of exotic butterflies by Geyer which are not figured in works by Hübner or Herrich-Schäffer. Includes notes by Herrich-Schäffer.

File 16: Drawings of *Papilio* by Geyer, annotated by Herrich-Schäffer.

File 17: Drawings of exotic Danaidae, Satyridae and Ithomiinae by Geyer.

File 18: Drawings of exotic Nymphalidae by Geyer.

File 19: Numbered file, but no contents.

File 20: Set of wrappers for the 1834 part of the *Sammlung europäischer Schmetterlinge.*

File 21: Drawings of Erycinidae, Lycaenidae and Noctuidae by Geyer.

File 22: Drawings of butterflies, possibly used in Herrich-Schäffer's *Systematische Bearbeitung der Schmetterlinge von Europa* (1843-1856).

File 23: Drawings of moths, as in File 22.

File 24: Drawings of European Microlepidoptera by Herrich-Schäffer.

File 25:
- (i) One notebook containing notes and drawings.
- (ii) Two copies of Herrich-Schäffer's *Prodromus Systematis Lepidopterorum* (1864-1869).
- (iii) A photograph of Herrich-Schäffer.
- (iv) 30 letters from Carol Plötz (1813-1886), dated 1866-1871.

File 26: Drawings *ex libris* Herrich-Schäffer.

File 27: Drawings of Lepidoptera attributed to Geyer, some possibly by Herrich-Schäffer.

File 28: Drawings of Papilionidae and Lycaenidae, partly by Geyer and Plötz.

File 29: Drawings of Noctuidae and Geometridae by Herrich-Schäffer.

File 30: Drawings of Nymphalidae by Herrich-Schäffer.

File 31: Drawings of Danaidae by Herrich-Schäffer.

File 32: Drawings of Hesperiidae by Herrich Schäffer.

File 33: Drawings of Satyridae, Amathusiidae and Morphidae by Herrich-Schäffer.

File 34: Drawings of various families of moths by Herrich-Schäffer.

File 35: Incomplete set of wrappers for Herrich-Schäffer's *Systematische Bearbeitung der Schmetterlinge von Europa* (1843-1856).

File 36: Miscellaneous plates from Esper's *Die ausländischen oder die ausserhalb Europa . . . Schmetterlinge* ([1785]-1801).

File 37: Miscellaneous plates and text, possibly from Bertuch's *Picture Book for the Young*, n.d.

File 38: Incomplete set of the butterfly parts, from Panzer's *Faunae Insectorum Germanicae* ([1792-] 1793-1809[-1823]).

File 39: Incomplete set of the moth plates from Herrich-Schäffer's continuation of Panzer's *Deutschlands Insecten*, 1829-1844.

File 40: Three leaves of original drawings of moths by Joseph Mann, c. 1850.

File 41: Original drawings by Jakob Sturm (1771-1848), including four drawings for Sturm's *Insecten-Cabinet* (1791-1792).

File 42: Miscellaneous plates and cut-outs from Hübner's works.

File 43: Miscellaneous plates from works by Rosel, Hübner, Herrich-Schäffer and others.

File 44: Anonymous drawings and manuscript notes on butterflies and caterpillars.

File 45: Proof plates by Carl Wilhelm Hahn (d. 1836) and C.F. Freyer.

Files 46-47: Numbered files, but no contents.

File 48: Original drawings of Acraeidae and Hesperiidae, by Carl Plötz for Herrich-Schäffer.

File 49: Original drawings of Sphingidae by Plötz .

File 50: Original drawings of Erycinidae and Pieridae by Plötz .

Files 51-54: Numbered files, but no contents.
File 55: Proof plates from various works by Hübner and Herrich-Schäffer.
File 56: Annotated original drawings of Microlepidoptera by Herrich-Schäffer.
File 57: Drawings by Geyer, copied from paintings by John Abbot and possibly C. Stoll.
File 58: Copies of drawings attributed to Plötz.
File 59: Manuscript by Herrich-Schäffer, forming a supplement to the *Fauna Ratisbonensis*, 1840.
File 60: Tracings by Geyer.
File 61: Miscellaneous notes by Herrich-Schäffer.
File 62: Numbered file, but no contents.
File 63: Proof plates from various works by Hübner.

280. ROTHNEY, George Alexander James (1849-1922)

He formed an extensive collection of Oriental Hymenoptera during his travels in India. His library and collection were donated to the Hope Entomological Collections, Oxford.

Biographical Reference: 1922. Anon. *Entomological News* **33**:255.

Collection

(1) One loose-leaf manuscript: [Aculeate Hymenoptera collected in Barrakpore Park, India], 1872-1886, 1893.

281. ROTHSCHILD, Nathaniel Charles (1877-1923)

Graduating from Trinity College in 1898, he entered the firm of Messrs N.M. Rothschild & Co. His interest in Natural History concentrated on the ectoparasites of mammals and birds. He described many new species and contributed papers to journals including *Novitates Zoologicae*. Working with Karl Jordan, he discovered *Xenopsylla cheopis*, the rat flea, and chief transmitter of bubonic plague in Egypt. His large collection of ectoparasites was presented to The Natural History Museum, London. N.C. Rothschild supported the work of the Imperial Bureau of Entomology and the Society for the Protection of Nature Reserves. He was a Fellow of the Entomological Society of London and was its President, 1915-1916. He was also a Fellow of the Linnean and Zoological Societies.

Biographical Reference: 1923. Walker, J.J. *The Entomologist's Monthly Magazine* **59**:279-280.

Collection

(1) Six Ms leaves: *Lepidoptera of the Fens in the neighbourhood of Cambridge.*
(2) Annotated Ts copy: *The Breck.*
(3) 15 original drawings of moths, with lithographed plates by W. Purkiss.
(4) Five photographs of the Fens in 1900s.
(5) Ts: *Books and articles on the Fens.*
(6) Two copies of an engraved illustration of a butterfly collector (possibly Rothschild).

282. SALLÉ, Auguste (1820-1896)

A French entomologist who travelled throughout the United States, West Indies, Central America and Venezuela making large collections of natural history specimens, especially insects. He then returned to Europe and established himself as a natural history agent in Paris. F.D.Godman and O. Salvin made use of Sallé's Central American collections in the compilation of *Biologia Centrali Americana* (1879-1915). He was a Fellow of the Société Entomologique de France, and of the Entomological Society of London.

Biographical Reference: 1896. Anon. *The Entomologist's Monthly Magazine* **32**:141.

Collection

(1) One loose-leaf Ms catalogue: *The Sallé collection of Coleoptera from the library of George Charles Champion (1851-1927).*

283. SAMOUELLE, George (d. 1846)

As a young man Samouelle trained to be a bookseller. He became interested in entomology and amassed a large collection of British insects. In 1819 he published *The Entomologist's Useful Compendium,* and in 1833-1834 *The Entomological Cabinet*. In 1821 he was appointed assistant to Dr William Elford Leach (1790-1836) in the Natural History Department of the British Museum. Samouelle had a chequered career at the Museum. He was suspended for a few months in 1838 as a consequence of putting himself forward as security for his son-in-law's debts. In 1841 he was dismissed following complaints about his conduct. During his time at the Museum he kept careful notes and relevant manuscripts about the collections, which are now bound in three volumes entitled *Entomological Memorandums*.

Collection

Entomological Memorandums:
Volume 1: 1838-1840
(1) General register of insect specimens presented and purchased 1830-1840.
(2) One Ms leaf: *Insects exchanged with J.O. Westwood.*
(3) One Ms leaf: *Insects from Greenland and Hamburgh, presented by Mons. Boie of Kiel.*
(4) 26 Ms leaves: *List of Lepidoptera purchased of M. Becker.*
(5) Four leaves: [Priced catalogue of German Lepidoptera – Ms annotations] and *Verzeichniss der Schmetterlinge der Gegend um Braunschweig* .
(6) Eight Ms leaves: *Insects from Columbia April 30 1840, with their prices in French money.*
Volume 2: 1822-1840
The following items are listed in the order in which they are bound . The Mss are interleaved with copies of internal memoranda from Samouelle to the Trustees regarding specimens.
(1) Two Ms leaves: *Department of Natural History. Report of the state of the collection of insects contained in 15 cabinets of 40 drawers each.* G.S. n.d, pp.3-4.
(2) One Ms leaf: *Crustacea and insects named and inserted in the 'General Zoological Catalogue' of the British Museum, 8 May 1833*, p.28.

(3) One Ms leaf: *State of the Crustacea cabinets British Collection in 4 cabinets of 40 drawers each.* 1833, pp.29-30.

(4) Four Ms leaves: *Insects etc Purchased and Presented - 1835,* pp.73-74.

(5) Copies of Memoranda from G. Samouelle to the Trustees regarding the numbers of specimens in the collections.

(6) One Ms leaf: *Notes on Coleoptera by [A.] Chevrolat,* p.214.

(7) One Ms leaf: *Insects selected from Box no.6 Java and Columbia. Purchased by Mr. Warwick* May 1839, pp.215-218.

(8) One Ms leaf: *List of the subjects sent by Mr Redman,* 1824, p.219

(9) Four Ms leaves: *Insects from Siberia collected by Gether and presented by Dr. Leach to be numbered and placed in Cabinets,* 1823, p.223

(10) One Ms leaf: *567 specimens of British Insects many of them new to this country presented by Dr. W.E. Leach,* 1826, p.226.

(11) One Ms leaf: *A list of British insects for the British Museum,* J. C. Dale, 1824, p.231.

(12) 32 Ms leaves: *British insects presented by Francis Walker Esq,* 1834-1835, pp.233-266.

(13) Two Ms leaves: *Insects presented by the Revd D.F. Morgan.* 1835, pp.267-268.

(14) Three Ms leaves: *Insects bought by Mrs. Henry,* 1835, pp.270-272.

(15) 24 Ms leaves: *Hardwicke Bequest 1835. Insects. Order Coleoptera,.* pp.273-298.

(16) One Ms leaf: *Insects from the Navigators Islands, presented by His Grace The Duke of Northumberland,* p.299.

(17) 16 Ms leaves: *Insects presented by Francis Walker Esqr.,* 1837-1838, p.300-315.

(18) Two Ms leaves: 1836. *Presents and accessions to the Entomological Collections and when the additions have been catalogued etc.,* pp.316-317.

(19) One Ms leaf: *Diptera from Mr. Horsfield in East India Museum,* p.319.

(20) Ten Ms leaves: *Crustaceorum quod in Museo Berolinense Conservantur. Catalogus,* pp.320-332.

(21) Two Ms leaves: *Alphabetic note of my Desiderata of exotic Lepidoptera named after Cramer,* Henry Escher Zollikofer, 1822, pp.337-338.

(22) Two Ms leaves: Letter from De Brebisson to J. Children, pp.343-344.

(23) Five Ms leaves: *Catalogue of Carabids,* [Reich Collection] and one letter to Samouelle from Hemmerith, pp.351-353.

(24) Three Ms leaves: *Doublets of German insects classed after the system of Latreille*, p.356-358.

(25) One Ms leaf: *Prospectus d'un ouvrage dessiné, contenant des Diptères ou Insectes a deux ailes indigènes en Europe.* J.C. Meigen, 1826, p.367.

(26) One Ms leaf: *Insectes de la Toscane. Hommage de Charles Passarini au Museum de la Société Royale de Londres (3 mars 1837)*, p.369.

Volume 3: 1819-1836

(1) Lists of insects received in The British Museum:

- (i) *From Prof. Hellwig BM 1821-40*, p.1.
- *(ii)* *From Siberia Collected by D.F. Gebler, 1819*, p.2.
- (iii) *Insects bought of J. Sowerby, 1820*, p.3.
- (iv) *Insects from South America bought of Mr. Bullock, 1820*, p.4.
- (v) *Insects received from Capt. Redman in 1820 collected in N. America*, p.5.
- (vi) *Insects from China, presented by Edmund Larkin, 1824*, p.7.
- (vii) *Insects from Demarara Bt. of M. Bowers & Co, 1824*, p.8.
- (viii) *Crabs and insects sent to Mr. Burchell in exchange for African insects*, p.9-11.
- (ix) *From Brazil presented by the Rev. J.W. Worthington, 1824*, p.12
- (x) *Insects from the Isle of France bought by Mr. Marve, 1825*, p.13.
- (xi) *Insects from New Holland collected during a survey of the North West Coast by J. Hunter, 1825*, p.14.
- (xii) *Insects from South America presented by Sir Charles Stewart, 1826*, p.15.
- (xiii) *Insects from Bengal presented by Major General Harwicke, 1826*, p.16.
- (xiv) *Insects from Chile, Brazil and the Sandwich Is., collected by Lord Byron, 1826*, p.17.
- (xv) *Specimens of European Insects presented by W.E. Leach, 1826*, pp.18-20.
- (xvi) *British insects taken in Devonshire by W.E. Leach during the Autumn of 1825 and the Summer of 1826*, pp.21-25.
- (xvii) *Insects from Madeira Recd. from Dr. Heinekin, 1828*, p.26.
- (xviii) *Insects from Columbia presented by Capt. Friend, 1828*, p.29.
- (xix) *Egyptian Insects presented by James Burton, 1828*, pp.30-32.

(xx) *Rare British Insects presented by J.G. Children, 1829*, p.33.

(xxi) *British Insects presented by J.C. Dale, 1829*, p.34.

(xxii) *List of the Lepidopterous insects selected for the British Museum from the Collections of the Hon. East India Company (in Dr. Horsfield's hand)*, pp.35-36.

(xxiii) *Insects presented by the Lord Coss of the Admiralty from H.M. ship The Chanticlear, 1831*, pp.39-41.

(xxiv) *Insects presented by Mr. Guizy, 1831*, p.42.

(xxv) *Insects Br. of Mrs. Noel 1831*, pp.43-44.

(xxvi) *Insects from N. America presented by J.G. Children, 1832*, pp.45-46.

(xxvii) *Insects from China presented by J. Reeves, 1832*, p.47.

(xxviii) *Insects Bought at Mr. Lees sale, 1832*, pp.48-49.

(xxix) *Insects Br. of Mrs. Noel, 1832*, pp.50-51.

(xxx) *Verzeichniss einiger vom Ludwig Parreyss aus Wien übernommenen Coleoptera*, pp.52-59.

(xxxi) *Insects presented in November 1832 from Siberia by M. Grey Esq*, p.60

(xxxii) *New Holland* [Australia] *Insects Br. of Mr. Stutchburg, 1833*, p.61.

(xxxiii) *Insects from Mexico presented by J. Taylor Esq. of Bedford Row, 1833*, p.62.

(xxxiv) *Insects collected between Senegal and Gambia River and presented by John D. Tebbs, 1833*, pp.63-65.

(xxxv) [Insects presented by F.Walker], p.67-68.

(xxxvi) *57 specimens of New Holland insects recd. from Alex Macleay, 1833*, p.66.

(xxxvii) *Specimens of insects from Senegal Bt. of Mr. John Curtis for the sum of £15.0.0. 1834*, pp.69-74.

(xxxviii) *Major G. Hardwicke's collections received at the Museum April 9, 1835*, pp.76-78.

(xxxix) *Rev. D.F. Morgan's African Insects, 1835*, p.79.

(xl) *Insects presented by Rev. D.F. Morgan, 1835*, pp.80-83.

(xli) [Insects presented by F.Walker] 1836, p.97-98.

(xlii) *African Insects Br. of Mr. John Curtis 1836*, pp.92-93.

(xliii) *Mexican Insects Br. of Mr. J. Curtis*, pp.94-96.

(xliv) *Insects from the Himalaya Presented by Lady S. Amherst, 1836*, p.99.

(xlv) [Insects presented by F.Walker] 1830, p.100.

(xlvi) *Coleoptera Br. Mr. Sowerby by Dr.Leach, 1820*, pp.101-105.

(xlvii) Coleoptera from New Holland presented Nov. 1821 by Lord Norris, p.106.

(xlviii) *Siberian Coleoptera collected by Gebler*, 1822, p.107.

(xlix) *Brasilian Coleoptera presented by The Rev. J.W. Worthington, 1824*, p.201.

(l) *Brasilian Insects brt. of Mr.Bullock by Dr. Leach, 1820*, p.205.

(li) *List [of insects] sent by Dr. Heineken of Madeira*, pp.237-246.

(lii) *List of Insects and crustacea sent to* [P.A] *Latreille*, pp.247-248.

(2) *Observations made on Dr. Billberg's Collection*, pp.250-271.

(3) *Copy of paper regarding nest of* Myrapetra scutellaris *sent by W. Hawkins July 1, 1843*, p.272.

(4) 15 Ms leaves: *A Report of Entomological Cabinets, Boxes, etc., in the Insect Room 1840*, pp.273-287.

284. SAMPSON, Frank Winn (1853-1926)

Sampson spent many years as a government offical in Nigeria. His private interest was the study of Scolytid beetles and he published papers mainly in *The Annals and Magazine of Natural History*. His entomological collection and library were bequeathed to The Natural History Museum, London. B.M.1926-428. He was elected a Fellow of the Entomological Society of London in 1906.

Biographical Reference: 1927 Anon. *The Entomologist's Monthly Magazine* **63**:16.

Collection

(1) One loose-leaf file: *F.W. Sampson. Catalogue of Platypodidae.*

(2) One Ms. notebook: *Catalogue (slides).* [Generic index to slide collection, mostly of Coleoptera, presented to The Natural History Museum, London]. B.M.1926-482.

285. SAUNDBY, Robert Henry Magnus Spencer (1896-1971)

Saundby served over 30 years in the Royal Air Force, joining The Royal Flying Corps in 1916. During World War II he was Deputy Air Officer Commander-in-Chief of Bomber Command Headquarters. His particular interest in natural history concerned the geographical variation of species of British Lepidoptera. He was President of the British Entomological and Natural History Society and Deputy Curator of Newbury Borough Museum. His collection of British Lepidoptera was bequeathed to The Natural History Museum, London. B.M.1971-594.

Biographical Reference: 1972. B.R.B. *Proceedings and Transactions of the British Entomological and Natural History Society* **5**:31-32.

Collection

(1) Four entomological diaries, 1943-1971.
(2) One Ms notebook:[Lepidoptera captures].
(3) One loose-leaf Ms: *Various statistics and records.*
(4) One Ms notebook: *Setting Log from 1964-1971.*

286. SAUNDERS, Charles James (1868-1941)

Educated at St. John's College, Oxford, Saunders went to Singapore in 1891 to serve in the Colonial Civil Service. He formed a collection of Beetles that was presented to The Natural History Museum, London and his Hemiptera, Heteroptera to the Lewes Museum.

Biographical Reference: 1941. Blair, K.G. *The Entomologist's Monthly Magazine* **77**:209.

Collection

(1) Three Ms volumes: [Register of Saunders collection.]
(2) One collecting notebook.
(3) One Ms notebook: *Host plants of British Beetles.*
(4) Ten letters from R. Hanitsch, 1928-1939.

287. SAUNDERS, Edward (1848-1910)

Saunders studied the Buprestidae of the world and visited all the main European museums which had collections of Buprestid type material. In 1871 he published *Catalogus Buprestidarum synonymicus et systematicus*. Later he studied and published papers on British Hemiptera and the Aculeate Hymenoptera. In 1892 he published, *The Hemiptera Heteroptera of the British Islands*, and in 1896, *The Hymenoptera Aculeata of the British Isles*. He was a Fellow of the Entomological Society, the Linnean Society and the Royal Society.

Biographical Reference: 1910. Anon. *Entomologist's Record and Journal of Variation* **22**:75-76.

Collection

(1) One Ms notebook: *Notes on Buprestidae*. This notebook contains notes on Buprestidae found in collections during his European tour.
(2) One Ms notebook: *Register of Buprestidae*, 1872.

288. SCHARFF, J.W.

Collection

(1) One Ms notebook: [Collecting activities in Malaya] 1919-1920.

289. SCHMOLLE, J.

The Schmolle collection of Lepidoptera from Britain and Africa was donated to The Natural History Museum, London in 1972. B.M.1972-274.

Collection

(1) One Ms notebook: [British and African Lepidoptera collection.]

290. SCHWANWITSCH, B.N.

Collection

(1) Ts of a paper: *Studies upon the wing-pattern of Catagramma and related genera of South American Nymphalid butterflies* (1929), *Transactions of the Zoological Society of London* **21**:105-294. Also includes proofs of text and figures.

291. SCOTT, F.B.

Scott was a Lieutenant-Colonel in the Indian Army. With Thomas R.D. Bell he bred many Indian hawk moths and made careful descriptions of their early stages, life-histories and habits. Scott and Bell were authors of the *Sphingidae* volume (1937) in the series, *The Fauna of British India*. Scott was a Fellow of the Royal Entomological Society of London.

Collection

(1) Two boxes Ms: *Fauna of British India Sphingidae* (1937).
(2) Five Tss:
  (i) *Notes on the* Brahmea wallichii, *larvae and moth.*
  (ii) *Camping in Central India.*
  (iii) *A season in Assam Hill Tracts.*
  (iv) *Camping in Assam.*
  (v) *Notes on a Red Sea Island.*
(3) Correspondence includes letters relating to *The Fauna of British India, Sphingidae* (1937) 1924-1936. Includes letters from:

Bell, Thomas Reid Davys (1863-1948)	1924-1934
Edwards, Frederick Wallace (1888-1940)	1934
Talbot, George (1882-1952)	1936
Tams, Willie Horace Thomas (1891-1980)	1934

292. SCOTT, Hugh (1885-1960)

Scott studied natural science at Trinity College, Cambridge. In 1908 he joined the Second Percy Sladen Trust Expedition, studying the fauna of the

Seychelles Islands, and was the general editor of the insect parts of the Seychelles reports. On his return to England in 1909 he became the Curator of Entomology at Cambridge University Museum and held this post until 1928. In 1930 he became Assistant Keeper in the Department of Entomology of The Natural History Museum, London, and held this post until his retirement in 1948. His publications included many scientific and geographical papers, and in 1942 he published *In the High Yemen*. He was a Fellow of the Royal Entomological Society, London.

Biographical Reference: 1960. Benson, R.B. *The Entomologist's Monthly Magazine* **96**:105, portrait.

Collection

(1) One Ms notebook:
- (i) *List of localities of insects obtained in the Seychelles Islands between July 1908 and March 1909.*
- (ii) *List of localities of insects obtained in Aldabra* [Indian Ocean] *Assumption and Cosmoledo Islands* [Seychelles]. *August 1908-February 1909 by Mr. J.C.F. Fryer.*
- (iii) *H.M.S. Sealark*,1905.

293. SELANGOR MUSEUM

Collection

(1) One bound Ts: *A catalogue of the Selangor Museum, Kuala Lumpur, Malaya. Lepidoptera, Rhopalocera, Papilionidae 1948.* A note in the volume states that many specimens were missing from the collection on arrival at The Natural History Museum, London, 31.8.1955.

294. SEVASTOPULO, Demetrius George (1903-1987)

Born in London, he was educated at Haileybury College, and later lived in Mombasa, Kenya. His main interest was the Lepidoptera and he published more than 300 papers in journals, including *The Entomologist*, *Transactions of the Royal Entomological Society of London* and the *Transactions and Proceedings of the South London Entomological and Natural History Society.*

Biographical Reference: 1987. Anon. *News of the Lepidopterists' Society,* **5**:74 .

Collection

(1) One loose-leaf Ts: *The early stages of Indian Lepidoptera,* c. 1950.
(2) Four boxes of loose-leaf Mss and 12 bound volumes: *The early stages of East African Lepidoptera,* including many black and white photographs, c. 1940s-1950s.

295. SHACKLETON-ROWETT EXPEDITION, 1921-1922

John Quiller Rowett provided the financial backing in 1921 for an expedition led by Sir Ernest Henry Shackleton to observe and collect scientific data in the Antarctic and sub-Antarctic areas; this included the collection of natural history specimens. During the expedition Shackleton died and was buried in South Georgia. Insect specimens from the expedition were donated to The Natural History Museum, London. B.M.1922-363.

Biographical Reference: 1923. F. Wild and A.H. Macklin. *Shackleton's Last Voyage. The Story of the Quest.* xvi + 373pp. London.

Collection

(1) Loose-leaf Ms: [Reports of Insect Specimens collected from the Shackleton-Rowett Expedition to the Antartic].

296. SHARP, David. (1840-1922)

Sharp studied and qualified for a medical career. In 1890 he became curator of the University Museum of Zoology at Cambridge. He became interested in Coleoptera and published over 250 papers. He was a Fellow of the Linnean Society, Zoological Society, Royal Society and Entomological Society of London, serving the latter as Vice-President, President and Secretary. His collection of Coleoptera was donated to The Natural History Museum, London and his entomological library went to the Cawthron Institute, New Zealand.

Biographical References: 1922. J.J.W. *The Entomologist's Monthly Magazine* **58**:234-237; 1924. Perkins, R.C.L. *Proceedings of the Royal Society of London* (B) **95**: xxxvi-xlii, bibliography, portrait.

Collection

(1) One Ms notebook: *Dr. David Sharp Collection of British Coleoptera: Catalogue of localities*, 1861-1871. [Includes some explanatory notes by Hugh Scott, 1922.]
(2) One bound Ts: [Index to David Sharp Coleoptera Collection.]

297. SHELDON, William George (1859-1943)

Sheldon was a keen lepidopterist, with a particular interest in the Tortricidae. He travelled widely in Europe, collecting Lepidoptera. He was treasurer of the Royal Entomological Society of London.

Biographical Reference: 1944. Riley, N.D. *The Entomologist*. **77**:17-19.

Collection

(1) Two Ms volumes: [Collecting notes and data].
(2) One Ms volume: *Notes of Life Histories*.

298. SHELJUZHKO, Leo (1890-1969)

Born in Kiev, he studied Natural Sciences at Kiev University. After graduating he bred exotic fish and water plants and this became an established business. Sheljuzhko concentrated on his lifelong interest in butterflies and built up a major collection of butterflies from the Palaearctic region. He corresponded with many lepidopterists and was a member of many entomological societies. After 1917 his personal circumstances changed and he handed his collection over to the Zoological Museum in Kiev. Following World War II he started a new life in Germany, where he continued his entomological studies.

Biographical Reference: 1971: W.F. *Nachrichtenblatt Bayerischen Entomologen* **20**:81-86.

Collection

(1) One Ms notebook: *Catalogus Lepidopterorum collectionis L. Sheljuzhko Pars I. Papilionidae.*
(2) One letter from Sheljuzhko to Karl Jordan, 1923.

299. SHELLEY

Collection

(1) Two Ms notebooks: Notes on butterflies, moths and birds [1879-1880].

300. SLOANE, Sir Hans (1660-1753)

Born in County Down, Ireland, he became one of the most significant scientific figures of the seventeenth and eighteenth centuries. On his return from an adventurous voyage to the West Indies, he established a successful medical practice. Hans Sloane became President of the Royal Society, 1727-1741, and the Royal College of Physicians, 1719-1735. His extensive collections were recognized to be of great significance, and on his death Parliament raised the money by lottery to purchase the collection for the nation. This formed the major part of the foundation of the British Museum. In 1881 the natural history collections were transferred to the new museum at South Kensington.

Biographical References: 1954. Brooks, E. St. J. *Sir Hans Sloane, The Great Collector and His Circle of Friends.* 234pp. London; 1994. Macgregor, A. *Sir Hans Sloane: Collector, Scientist, Antiquary, Founding Father of the British Museum.* 308pp. London.

Collection

(1) Two folio Ms volumes: Begins with 5 pages entitled *Tabula Synoptica Insectorum in Classes distributorum à Cromwello Mortimero. M.D-R.S. Secr,* by

Sloane's amanuenis, Cromwell Mortimer (1698-1752). This is followed by the main catalogue, *An Explanation of Several Abbreviations & Notes made use of in the Catalogue of Insects*, which refers mostly to individual insect specimens in the Sloane collection. These entries are in a number of hands, although the earlier entries are presumed to be by Sloane. The insect collection was an amalgum of collections belonging to such people as Joseph Dandridge and James Petiver. Another 15 catalogues of the Sloane collection exist in The Natural History Museum Libraries.

301. SMIT, Franciscus Gerardus Albertus Maria

A Dutchman by birth, Smit was appointed to the Zoological Museum at Tring to work on the Rothschild Flea Collection. He was a meticulous worker and published many papers on fleas and was the author of Volume 7 of *An Illustrated Catalogue of the Rothschild Collection of Fleas (Siphonaptera) in the British Museum (Natural History)* 1987. He was the originator and editor of *Flea News* until his retirement.

Collection

(1) Ms collection:
- (i) [Type data and literature on Siphonaptera.]
- (ii) Annotated Ts: *List of Siphonaptera not yet represented in the Tring Museum.*
- (iii) Draft Ms: *Siphonaptera from Hindu Kush*, 1973.
- (iv) [Notes and literature references to pathogens of fleas.]
- (v) [Notes and Mss of programmes for the International Conference on Fleas, 1977.]
- (vi) Ts: [Bibliography of Siphonaptera.]

(2) Notebooks:
- (i) [Sale of publications.]
- (ii) [Short list of flea names.]
- (iii) *Notes on Siphonaptera.*
- (iv) *Siphonaptera.*
- (v) [Flea notes, localities and copies of letters.]
- (vi) *Status of Siphonaptera Taxa.*

(vii) *Catalogue of Flea Collection at Tring, 1950-1958.*
(viii) *Species of Siphonatera with original references.*

302. SMITH, Frederick (1805-1879)

Born in London, Smith trained as a engraver on steel. He became interested in the habits of Hymenoptera. In 1850 he was appointed as Assistant in the Zoological Department of the British Museum and in 1875 became Senior Assistant. His published works include, *Catalogue of Hymenopterous Insects in the British Museum,* (1853-1859) and *Catalogue of the British Fossorial Hymenoptera* (1858). He was President of the Entomological Society of London in 1862 and 1863.

Biographical Reference: 1879. Anon. *The Entomologist's Monthly Magazine* **55**:263-264.

Collection

(1) Unpublished Ms: *New Hymenoptera.* Concerns the determination and arrangement of foreign Hymenoptera in the British Museum.
(2) Two Ms notebooks:
   (i) *Manuscript notes on British Insects* [includes original sketches].
   (ii) [Manuscript notes on Hymenoptera].

303. SOUTH, Richard (1846-1932)

As a young boy, South developed a love for natural history with a special interest in the Pyralidae, Tortricidae and Pterophoridae. In his late thirties, he started to contribute papers to entomological journals. In 1887 South began work with John Henry Leech, and as a result of this association he became interested in the Lepidoptera of the Far East, contributing to Leech's *Butterflies of China, Japan and Corea* (1892-1894). South's popular works, *Butterflies of the British Isles,* (1906) and the two-volume *Moths of the British Isles,* (1907-1908) became classic works, and were reprinted many times. South was elected Fellow of the Entomological Society of London in 1885 and joined the South London Entomological and Natural History Society in 1882.

He was employed by The Natural History Museum, London as a special assistant. He was the owner and editor of *The Entomologist* for many years.

Biographical References: 1932. A.E.S. *The Entomologist's Monthly Magazine.* **68**:118-119; 1932. Riley, N.D. *The Entomologist* **65**:97-100, portrait.

Collection

(1) One bound volume of correspondence: 16 letters to South, 1905-1913, relating to South's works, *Butterflies of the British Isles* (1906) and *Moths of the British Isles* (1907-1908). 12 letters to N.D. Riley, 1932, from various entomologists concerning new editions of South's works.

304. SOUTHGATE, Ben J. (d. 1985)

Southgate worked in the Pest Infestation Control Laboratory at Slough, studying Bruchid taxonomy. In later years he concentrated on detection and control of Bruchids in various stored grains.

Collection

(1) One Ms notebook: [Lists of the photographs of Bruchids in the C.J. Schoenherr and Maurice Pic collection]. Photographs referred to are held in the Entomology Department, The Natural History Museum, London.

305. SPENCE, William (1783-1859)

In 1805 Spence began his interest in entomology. He corresponded with William F. Kirby (1844-1912) and they became lifelong friends. In 1808 Spence proposed to Kirby a joint work and this resulted in the publication of the entomological classic, *An Introduction to Entomology* (1815-1826). After a period of ill health, Spence left England in 1826 for a tour of European capitals. In 1847 Spence was elected President of the Entomological Society of London.

Biographical Reference: 1861. Stainton, H.T. *Proceedings of the Entomological Society,* **1858-61**:92-94.

Collection

(1) One bound volume: 12 letters signed by Spence, dated 1807-1857, concerning *An Introduction to Entomology* (1815-1826).

306. SPOONER, Malcolm (1907-1989)

A marine biologist by profession, his career was with the Marine Biological Association, Plymouth. He was awarded an MBE for his work during World War II at Bletchley Park, helping to break the German codes. His particular interest was British Aculeate Hymenoptera. He published 38 papers on British Aculeates and Diptera. His collection of several thousand specimens was given to The Natural History Museum, London. B.M.1986-349. Specimens were also given to the University Museum of Zoology, Cambridge and the University Museum, Oxford.

Biographical Reference: 1990. Else, G.R. *The Entomologist's Monthly Magazine* **126**:173-176, bibliography, portrait.

Collection

(1) 40 field notebooks, 1926-1986.
(2) Seven miscellaneous notebooks including:
  (i) *Devonshire Aculeates.*
  (ii) *Moths (macro) recorded at Yelverton at light (mainly) or in house; also at the sap of an infested oak tree.*
  (iii) *Butterflies up to 1972.*

307. STAINTON, Henry Tibbats (1822-1892)

Stainton was educated first privately and them at Kings' College, London. He travelled extensively in Europe and corresponded with many European entomologists. Although he had a general interest in natural history, his main area of study was the Microlepidoptera. He contributed many papers to *The Zoologist*, established *The Entomologist's Annual* and helped to found *The Entomologist's Monthly Magazine*. His publications include, *The Entomologist's*

Companion (1851-1852), *A Manual of British Butterflies and Moths* (1857-1859) and *A Natural History of the Tineina*, published in 13 volumes (1855-1873). For the latter work he received assistance from P.C. Zeller, J.W. Douglas and H. Frey. Stainton was a Fellow of the Entomological Society of London and was President in 1881-1882. He was a Fellow of the Linnean Society, the Royal Society and a member of several entomological societies in Europe.

Biographical Reference: 1893. Douglas, J.W. *et al. The Entomologist's Monthly Magazine* **29**:1-4, bibliography, portrait.

Collection

(1) Two Ms vols: *Catalogue of British Lepidoptera in Coll. H.T. Stainton.*
(2) One Ms collecting notebook, 1851-1859.
(3) One copy of *A Manual of British Butterflies and Moths* (1857-1859), annotated by John Hartley Durrant.
(4) (i) Annotated copy: *List of British Tineina for interchange among collectors*, 1854.
(ii) Annotated copy: *List of Tineina larvae figured and described for the Natural History of Tineina*, 1863.
(iii) Loose-leaf Ms: *Descriptions of British larvae used for a Natural History of the Tineina (1855-1873) with notes by John Hartley Durrant and Lord Thomas Walsingham.*
(5) Loose-leaf Ms: *Ms relating to Stainton's collection of drawings prepared for A Natural History of Tineina* (1855-1873), by John Hartley Durrant.
(6) Eight leaf mines collected by Baron von Nolcken, from St. Thomas and Santa Marta, West Indies, 1870. Drawings of these leaves by G. Shulz.
(7) Loose-leaf Ms: [Descriptions of Microlepidoptera collated with the figures and arranged in the sequence of Staudinger's Catalogue by J.H. Durrant.]
(8) Proof plates for *A Natural History of Tineina* (1855-1873).
(9) Loose-leaf Ms: *Mr. H.T. Stainton's Collection of Drawings Illustrating the Life-history of Micro-Lepidoptera. A list exhibiting the exact sequence of the drawings when they were handed over to the British Museum*, by J.H. Durrant, 1893.
(10) One Ms notebook: *Tabulations of Stainton manuscripts and figures*, by J.H. Durrant.

(11) Loose-leaf Ms: *Syrian Noctuae* by J. Lederer January 1867.

(12) Four Ms notebooks: *Registers of Specimens* and 13 notebooks, 1848-1890.

(13) Two copies of Stainton's *Novitates Staintonianae or Catalogue of British Lepidoptera added to the Collection of H.T. Stainton, Esquire in the year 1842.* Includes 6 plates of original watercolour drawings. Twenty copies were circulated in the author's copperplate autograph.

(14) Twelve boxes of approximately 3,000 drawings by the artist William Wing (1827-1855) and his daughter M.L. Wing for *A Natural History of Tineina* (1863).

(15) Five letterbooks containing copies of letters to and from Philipp Charles Zeller, 1848-1856.

(16) Correspondence collection of over 2,000 letters from foreign entomologists to Stainton, c. 1850-1892, including letters from:

Belfrage, Gustaf Wilhelm (1834-1882)	1868-1874
Boheman, Carl Heinrich (1796-1868)	1854-1856
Clemens, James Brackenridge (1829-1867)	1857-1863
Constant, Alexandre (1829-1901)	1860-1889
Fernald, Charles Henry (1838-1921)	1879-1889
Fologne, Egide (d. 1919)	1856-1869
Herrich–Schäffer, Gottlieb August Wilhelm (1799-1874)	1850-1874
Frey, Heinrich (1822-1890)	1855-1890
Hofmann, Ernst (d. 1892)	1863-1891
Hofmann, Ottmar (1835-1900)	1857-1889
Kock, Gabreil (1807-1881)	1851-1860
Lederer, Julius (1821-1870)	1862-1865
Mann, Joseph Johann (1804-1889)	1849-1880
Mühlig, Johann Gottfried Gottlieb (1813-1884)	1855-1879
Nolcken, Baron von	1867-1884
Ragonot, Emile Louis (1843-1895)	1871-1888
Schmid, Anton (1809-1899)	1855-1892
Staudinger, Otto (1830-1900)	1853-1888

(17) Correspondence collection of over 11,000 letters from British entomologists and naturalists to Stainton, c. 1852-1892. Includes letters from:

Baker, G.T.	1859-1892
Barrett, Charles Golding (1836-1904)	1854-1892
Buckler, William (1814-1884)	1857-1884
Burdon, H.	1864-1884
Darwin, Charles Robert (1809-1882)	
(copies; originals held in General Library)	1860-1881
Doubleday, Henry (1809-1875)	1852-1875
Douglas, John William (1814-1905)	1853-1890
Dunning, Joseph William (1833-1897)	1852-1886
Grey, J.E.	1854-1871
Hellins, John (1829-1887)	1857-1887
Herbert, C.	1881-1887
Hodgkinson, James B. (1823-1897)	1856-1886
Jackson, W.H.	1872-1879
Janson, Edward Wesley (1822-1891)	1858-1873
Jeffrey, W.R.	1857-1886
Jordan, Robert Coane Roberts (1825-1890)	1852-1890
Kippist, R.	1862-1878
Kirby, William Forsell (1844-1912)	1862-1892
Knaggs, Henry Guard (1832-1908)	1862-1877
Legge, A.	1874-1889
Legge, H.	1866-1884
Logan, R.F.	1852-1855
Lubbock, John William. Lord Avebury (1834-1913)	1855-1881
McLachlan, Robert (1837-1904)	1860-1887
Rye, Edward Caldwell (1832-1885)	1860-1879
Sang, John (1828-1887)	1853-1886
Saunders, Edward (1848-1910)	1853-1892
Scott, John (1828-1888)	1853-1884
Wailes, George (1802-1882)	1854-1870
Walsingham, Thomas de Grey, (1843-1919)	1857-1888
Wing, William (1827-1855)	1852-1854
Wing, M.L.	1857-1886
Wollaston, Thomas Vernon (1822-1878)	1856-1874
Wood, John Henry (1841-1914)	1885-1891

(18) One Ms notebook: [*Records of identifications of North American Lepidoptera*] by J.H. Durrant 1881.
(19) Second series of correspondence from British entomologists, includes approximately 400 letters, c. 1865-1880, including letters from:

Buckler, William (1814-1884)	1869-1880
Douglas, John William (1814-1905)	1880-1896
Eaton, Alfred Edwin (1845-1929)	1880
Hodgkinson, James B. (1823-1897)	1880
McLachlan, Robert (1837-1904)	1880
Millière, Pierre (1811-1887)	1868-1879
Smith, Frederick (1805-1879)	1858-1864
Westwood, John Obadiah (1805-1893)	1856-1881

308. STEELE, Mary

Mary Steele collected insects whilst in Burma during the 1930s. She donated over 1,000 specimens to The Natural History Museum, London. B.M. 1937-324.

Collection

(1) One Ms notebook: [Locality and collecting data from Burma], 1936-1937.

309. STEPHENS, James Francis (1792-1852)

Stephens worked for many years as a clerk in the Admiralty, London. In 1818 he was seconded to the British Museum, and on completion of his work returned to the Admiralty. He made a large insect collection, which contained many type specimens. After his early retirement this collection was purchased by the British Museum. Stephens published many important works, including: *A Systematic Catalogue of British Insects* (1829), *Illustrations of British Entomology* (1827-46) in 12 volumes and *A Manual of British Beetles* (1839).

Biographical Reference: Westwood, J.O. *Proceedings of the Entomological Society of London* **1853**:45-50.

Collection

(1) Two Ms volumes: *Catalogue of the British Lepidoptera in Coll. J.F. Stephens.*

(2) Four packets of Mss:

Packet 1. *Mem's Entomological.*

- (i) Ms notes: [List of Linnean and English insect names.]
- (ii) *Catalogue of British Lepidoptera.*
- (iii) *Desiderata of Coleoptera.*

Packet 2. *Entomological Notes.*

- (i) Miscellaneous notes.
- (ii) Three Ms leaves by Thomas Marsham (d. 1819) for *Entomologia Britannica Coleoptera* (1802).
- (iii) *On Bees.*

Packet 3. *Nomenclature Notes.*

- (i) Miscellaneous notes.
- (ii) One letter from Henry Doubleday (1809-1875) to Stephens, 1847.
- (iii) Author's annotated copy of Stephens' *Nomenclature of British Insects* (1829).

Packet 4. *Notes, Entomological.*

- (i) One Ms: [Stephens' reminiscences.]
- (ii) 24 letters from Henry Doubleday to Stephens, 1842-1846.

310. STRINGER, Herbert

Stringer, a lepidopterist, was appointed by Lord Walsingham to help in the curation of his collection of Microlepidoptera. In 1910, as part of the Walsingham bequest, this large collection of 250,000 specimens was moved to The Natural History Museum, London. Stringer and John Hartley Durrant (1863-1928) continued to work on the collection at the Museum.

Collection

(1) One bound Ms: *List of Japanese, Formosan & Philippine Microlepidoptera*, 1926.

311. SWALE, Harold (1853-1919)

Trained in medicine, Swale developed a great interest in the Hymenoptera. In 1900 he went to New Zealand and spent three years collecting Coleoptera and Hymenoptera. On his return to England he decided to practise tropical medicine. He worked in the Sudan, India, East Africa and New Zealand. During his appointments he made extensive collections and he presented many specimens to The Natural History Museum, London over a period of years.

Biographical Reference: 1919. Andrewes, H.E. *The Entomologist's Monthly Magazine* **55**:140-141.

Collection

(1) Four Ms notebooks: [Hymenoptera notes], c. 1890s.

312. SWINHOE, Charles (1836-1923)

Swinhoe joined the army aged 19, serving in the Crimea and in India. While in India he collected Lepidoptera and sent many specimens to The Natural History Museum, London. He published several papers on Lepidoptera and contributed to Frederick Moore's *Lepidoptera Indica* (1890-1913). On the death of Moore in 1907, Swinhoe completed the work.

Biographical Reference: 1924. Riley, N.D. *The Entomologist* **57**:23-24.

Collection

(1) Two bound Ms volumes: *Butterflies* [a list of Swinhoe's butterfly collection].
(2) Loose-leaf Ms: Catalogue of the Swinhoe collection – Lycaenidae].
(3) Five Ms volumes: [Catalogue of the Swinhoe collection of oriental Lepidoptera].

313. TAMS, Willie Horace Thomas (1891-1980)

Born in Cambridge, Tams developed an early interest in natural history. He became assistant to Professor J.S. Gardiner at the University of Cambridge and in 1913 joined the staff of the Linnean Society. During World War I he served in the Canadian Army. In 1920 he was appointed to the staff of The Natural History Museum, London as Assistant Keeper in the Entomology Department. His collecting trips included the Percy Sladen and Godman Trust Expedition to the Gulf of Guinea, 1932-1933, Madeira and the Seychelles. Tams was a keen Lepidopterist and published over 70 papers on Lepidoptera taxonomy, including papers on the Lasiocampidae. He was a member of the British Entomological and Natural History Society, an associate of the Linnean Society and was assistant editor of the *Entomologist's Gazette*.

Biographical Reference: 1981. Fletcher, D.S. and Bradley, J.D. *Proceedings of the British Entomological and Natural History Society* **14**:34-40. Bibliography, portrait.

Collection

(1) One autograph book presented by staff of the Natural History Museum to Tams on his retirement.
(2) One Ms notebook and miscellaneous items.

314. TATUM, T.

Collection

(1) One Ms notebook: *Catalogue of Geodephaga Cabinets 1 & 2*, c. 1880. Refers to Tatum's insect collection, which was presented by John Charles Bowring (1821-1893) to the Natural History Museum, London. B.M.1863-47.

315. TAYLOR, J. Cleasley

Collection

(1) One letter from J.C. Taylor to the Director of the Natural History Museum, London, 15 January 1909, regarding locusts in Grand Canary. B.M. 1909-9.

Insect Room Lists: Volume 2, item 64

316. THEOBALD, Frederick Vincent (1868-1930)

Born in Kingston-upon-Thames, Theobald studied at St. John's College, Cambridge. He developed a keen interest in insects and especially economic entomology. He was much influenced by the work of Eleanor Ormerod (1828-1901). In 1894 he became a lecturer at the South Eastern Agricultural College in Wye and was later appointed Professor of Economic Entomology. He was a pioneer in this field of entomology. Theobald was also responsible for Economic Zoology at The Natural History Museum, London, and was invited by the Trustees to publish a work on the Culicidae of the world. For this work specimens were collected by Medical Officers around the world. This resulted in the publication, *A Monograph of the Culicidae, or Mosquitoes*, in 5 volumes, (1901-1910). After 1910 Theobald concentrated on agricultural entomology, publishing many papers including *The Plant Lice or Aphididae of Great Britain*, (1926-1929), in 3 volumes.

Biographical Reference: 1930. Laing, F. *The Entomologist's Monthly Magazine* **66**: 92-93, portrait.

Collection

(1) Correspondence collection: Approximately 900 letters from various entomologists on mosquito related matters, c. 1901-1903, including letters from:

Aitken, Edward Hamilton, (1851-1909)	1902-1903
Aldrich, John Merton (1866-1934)	1902

Bancroft, Thomas Lane (1860-1933)	1903
Blanchard, Raphael Anatole Emile (1857-1919)	1902-1903
Coquillett, Daniel William (1856-1911)	1903
Felt, Ephraim Porter (1868-1943)	1903
Froggatt, Walter Wilson (1858-1937)	1903
Goeldi, Emilio Augusto (1859-1917)	1901-1902
Green, Edward Ernest (1861-1949)	1901-1902
Herrick, Glenn Washington (1870-1965)	1903
Howard, Leland Ossian (1857-1950)	1902-1903
James, Sydney Price (1870-1946)	1903
Kellogg, Vernon Lyman (1867-1937)	1903
Ludlow, Clara Southmayd (1852-1924)	1901-1903
Lutz, Frank Eugene (1879-1943)	1902-1903
Ross, Ronald (1857-1932)	1902-1903
Sergent, Etienne (1878-1948)	1902
Wesche, Walter Francis Frederick (1857-1910)	1903

(2) Correspondence collection: Approximately 400 letters from Colonial Governers regarding the collecting of mosquitoes.

(3) Miscellaneous manuscripts.

(4) Queries sent to Theobald on economic zoology, c. 1899-1904.

(5) Ms, correspondence, proof plates and original drawings for Volume 3 (1903) of *A Monograph of the Culicidae, or Mosquitoes*.

317. THÉRY, André (1864-1947)

Born in France, but he spent much of his life in Rabat, Morocco. His main entomological interest was the Buprestidae and published many papers on this family of Coleoptera.

Biographical Reference: 1989. Bellamy, C.L. and Descarpentries, A. *Giornale Italiano di Entomologia* **4**:173-182, bibliography.

Collection

(1) Ms notes and drawings on the Buprestidae.

318. THORNLEY, Alfred (1855-1947)

Thornley graduated from Merton College, Oxford and became interested in natural history. During his later years he lived in Cornwall. He was a Fellow of the Royal Entomological Society of London for 56 years.

Biographical Reference: 1948. Williams, C. B. *Proceedings of the Royal Entomological Society, London* (C) **12**: 64.

Collection

(1) Loose-leaf Ms: [Cornish insect collecting notes], c. 1930s.

319. THURNALL, Alfred (1858-1929)

Employed as a draughtsman for the Great Eastern Railway Company, Thurnall developed a deep interest in the Tortricidae. From 1875 to 1929 he contributed over fifty notes and papers to *The Entomologist's Monthly Magazine*.

Biographical Reference: 1929. W. G. S. *The Entomologist* **62**:192.

Collection

(1) One collecting diary: [Lepidoptera], 1905. Annotation reads, 'This diary is a copy of my old one 1870-1905 with a few notes and corrections.'

320. TOMPKINS, Henry

Collection

(1) Nine Ms notebooks: Entomological collecting and specimen exchange notes. 1849-1850. One notebook also contains several editions of Stainton's *List of the British* Tinea *for interchange among collectors*.

321. TRIMEN, Roland (1840-1916)

In 1860 Trimen joined the Cape Colony Civil Service and studied the insect fauna of that area. In 1868 he read an important paper on the bionomics of *Papilio merope* to the Linnean Society. Other publications included the three-volume work, *South African Butterflies*, (1887-1889) by R. Trimen and J. H. Bowker. In 1873 he was appointed Curator of the South African Museum at Cape Town and remained there until his retirement. In 1883 he was elected a Fellow of the Royal Society.

Biographical References: 1916. Bethune-Baker, G. T. *Entomologist's Record and Journal of Variation* **28**:231-236; 1920. Poulton, E.B. *Proceedings of the Royal Society, London* (B) **91**:xviii-xxvii.

Collection

(1) One Ms notebook: *Roland Trimen's Collection of Extra-Tropical South-African Lepidoptera.* A complete record of collecting data, c. 1866-1867.
(2) One Ms notebook: *South African Butterflies. Trimen,* 1887.

322. TRING MUSEUM

Collection

(1) One Ms notebook: [Payments made for entomological collections received by Tring Museum, 1903-1938].

323. TROUGHT, Trevor (1891-1970)

Trought was employed by the Sudan Cotton Corporation and the Jordan Government. He was an authority on the genetics of cotton growing and was interested in the butterflies of India and the Middle East. In 1947, he was elected a Fellow of the Royal Entomological Society of London. His collections were presented to Warwick Museum.

Biographical Reference: 1971. Hinton, H. E. *Proceedings of the Royal Entomological Society of London* (C) **35**:54.

Collection

(1) One Ms notebook: [A general diary and travel notes], 1941.
(2) Two Ms notebooks: [Notes on Lepidoptera collecting in Jordan including localities], 1951-1955.

324. TURNBULL, L. D. J.

Lt.-Col. Turnbull worked for part of his life for the East Indian Railway as an engineer, based in Calcutta. He was an enthusiastic photographer with a keen interest in butterflies.

Collection

(1) One folder: 89 natural-size photographs of butterflies of India, 1945-1947, some handpainted by Turnbull.
(2) Approximately 100 loose photographs of butterflies. Some Ms annotations of localities.

325. TURNER, J.A.

Collection

(1) One Ms volume: notes attributed to Turner.
Bibliothèque de A.M. Labouchère, Index et Tableau comparatif des dessins originaux des Papillons exotiques de Cramer et des gravures contenues dans les quatre volumes de cet ouvrage c. 1780. [This is an autograph index to the published work by Pieter Cramer, *De uitlandsche Kapellen* ([1775] 1779-1782), and to the original drawings for this work which are held in the Entomology Library.]

326. TYTLER, Major General, Sir H.C. (1867-1939)

Born in Scotland, Tytler joined the army in 1886, serving in India, Europe, East Africa, Afghanistan and Burma, retiring in 1928. He presented The Natural History Museum, London, with insects from Burma; his collection of 2,500 Hesperiidae and 1,000 Lycaenidae from India was purchased in 1939. B.M.1939-614. In 1941 his collection of 23,000 Rhopalocera was purchased by the Natural History Museum, London from his widow. This collection included many type specimens.

Biographical Reference: 1941. *Who was Who* (1929-1940). London: Adam & Charles Black.

Collection

(1) Loose-leaf Ms notes: [The Tytler Indian Lepidoptera collection] by Sir Keith Cantlie.

327. UNWIN, William Charles (1811-1887)

Unwin had a wide interest in natural history, including ornithology, botany and entomology. His publications included a large number of papers, and in 1878 he published *Illustrations and dissections of the genera of British Mosses*. He lived for much of his life in Lewes, East Sussex.

Biographical Reference: 1887. Anon. *The Entomologist's Monthly Magazine* **24**: 7.

Collection

(1) One letter from Unwin to unknown person, 1846, includes a list of Lepidoptera and British birds found by Unwin in the neighbourhood of Lewes, East Sussex.

328. VERDOURT, Bernard

Collection

(1) One Ts: *The Insect Attack of Books*, written for the Printing Packaging and Allied Trades Research Association, 1948.

329. VINE, Arthur Charles (1844-1917)

Living in Brighton, Sussex, he collected Lepidoptera from the Downs between Brighton and Lewes and reared many obscure species of Microlepidoptera.

Biographical Reference: 1917. Adkin, R. *The Entomologist* **50**:40.

Collection

(1) One bound volume: 18 letters to Vine, 1878-1909, including 8 letters from Eustace Ralph Bankes (1861-1929), dated 1893-1907.

330. WALKER, Francis (1809-1874)

Born in Southgate, Middlesex , Walker developed an interest in entomology from an early age. He travelled widely, including visits to Lapland, Switzerland, France and Italy. He collected large numbers of specimens and supplied material to many entomologists including John Curtis and A.H. Haliday. He had a particular interest in the Chalcidoidea and in 1832 he published the introductory chapter of *Monographia Chalciditum* in the *Entomological Magazine*. In 1830 Walker and his brother, Henry toured Europe with John Curtis, who became a close friend. He also collected with the botanists William Christy junior and William Dawson Hooker on a voyage to Alten and the North Cape. In 1851 he published the first volume of three on Diptera in a projected series of works on British insects: *Insecta Britannica*. Other publications included a series of catalogues of insects in the British Museum collections, including the Lepidoptera, Diptera, Orthoptera, Homoptera and Neuroptera. He described many thousands of new species. Some of these descriptions were inadequate and these catalogues have dented his reputation as an entomologist.

Biographical Reference: 1874. Carrington, J.T. *The Entomologist's Monthly Magazine* **11**:140-141; 1874. Anon. *Entomologist* **7**:260-264.

Collection

(1) Loose-leaf Ms notes: [Diptera of Madeira].

331. WALKER, James John (1851-1939)

Walker served in the Royal Navy, and on his retirement in 1904 went to live in Oxford and collected local insects. In 1926 he published *The Natural History of the Oxford District* for the meeting of the British Association for the Advancement of Science. In the same year he joined the editorial staff of *The Entomologist's Monthly Magazine*. Walker was President of the Entomological Society of London and received an honorary degree from the University of Oxford. His collections came to The Natural History Museum, London in 1950, B.M.1950-548 and 1950-582.

Biographical Reference: 1939 K.G.B[lair].*The Entomologist* **52**:48.

Collection

(1) 15 Ms leaves: [Eight lists of insect collecting localities from around the world, written from HMS *Penguin*], c. 1890-1892.

Insect Room Lists: Volume 1, item 5.

(2) One Ms leaf: [List of various insects from around the world]

*nsect Room Lists:*Volume 2, item 67.

332. WALLACE, Alfred Russel (1823-1913)

Born in Usk, Monmouthshire, Wallace began collecting beetles with H.W. Bates in 1845, and in 1848 they travelled together along the River Amazon. Wallace published details of this expedition in *Travels on the Amazon and Rio Negro* (1853). In 1854-1862 he travelled around the Malay Archipelago and New Guinea, collecting over 100,000 insect specimens. In 1869 he published perhaps his most famous work, *Malay Archipelago: Land of the Orang-utan and the Bird of Paradise*. He published many papers in the journals of the Linnean, Zoological and Entomological Societies. In 1858 he sent his famous essay on species variation to Charles Darwin. Darwin's *On the Origin of Species* followed in 1859. Wallace was a Fellow of the Linnean, Zoological and Entomological Society of London.

Biographical Reference: 1913. Walker, J.J. *The Entomologist's Monthly Magazine* **49**:276-277.

Collection

(1) Correspondence collection: Approximately 100 letters to and from Wallace, 1900-1906, including letters from:

Godman, Frederick DuCane (1834-1919)	1903?
Jordan, Karl Hermann Christian (1875-1972)	1903
Kaye, William James (1875-1967)	1903-1905

Also approximately 30 letters to and from Frederick Birch, 1900-1906, regarding his expedition to British Guiana (Guyana).

333. WALSINGHAM, Thomas de Grey, Lord (1869-1919)

Born in London, he was the eldest son of Thomas, fifth Lord Walsingham and succeeded to the title in 1870. Educated at Eton and Cambridge, he then pursued a career in politics, but was an enthusiastic entomologist with a particular interest in Lepidoptera. He collected more than 50,000 specimens of Microlepidoptera in England, Europe, USA and Jamaica. He discovered the life-histories of many moths, and he described numerous new species. His British Lepidoptera collection included preserved larvae mounted in –natural positions on actual plants. Walsingham purchased microlepidoptera collections from Zeller, Hofmann, Christoph and others. His own collection was presented to The Natural History Museum, London in 1910, accompanied by his magnificent entomological library of books, manuscripts and correspondence. B.M.1910-427. He published many papers and monographs, including *Pterophoridae of California and Oregon* (1880). He was High Steward of the University of Cambridge, a Trustee of the British Museum (Natural History) and of the Hunterian Museum and the Lawes Agricultural Trust. He was President of the Entomological Society of London and a member of numerous other societies, including the Entomological Society of America, Société Entomologique de France and Nederlandse Entomologische Vereniging.

Biographical Reference: 1920. Durrant, J.H. *The Entomologist's Monthly Magazine* **56**:17, 25-28, portrait.

Collection

(1) One Ms leaf: Walsingham's first entomological note, aged 8 years. 'I have found out that the catipillars hind feete are different to its froant ones'.
(2) One bound Ts: *List of separata by Lord Walsingham and John Hartley Durrant* (1863-1928). The compiler is unknown (probably J.H. Durrant). Papers arranged in chronological order.
(3) Loose-leaf Ms: *The Walsingham Collection of British Lepidoptera and Larvae. A list of the specimens in the collection, and those needed to complete the collection.*
(4) Loose-leaf Ms: *New Pyralidina from the Punjab.* A Ms list of new Pyralidina from the Punjab in Lord Walsingham's collection, described by Alexander Benjamin van Medenbach de Roöy (1841-1878), 1875.
(5) Loose-leaf Ms: [Note on Pyralidae].
(6) Correspondence collection: letters from Walsingham, 1909-1911, concerning his gift of his Lepidoptera collection to The Natural History Museum, London B.M.1910-427.
(7) 45 numbered and bound volumes of manuscripts:

- (i) Volumes 1-7: List numbered specimens of various collections acquired and used by Walsingham, including Microlepidoptera from J.J. Walker.
- (ii) Volume 8: *Microlepidoptera received from Ernest Anderson,* Melbourne, Australia.
- (iii) Volume 9: [Collecting data from Hawaii, 1892; includes notes by R.C.L. Perkins].
- (iv) Volume 10: *Micros received from William G. Smith, taxidermist, Colorado, USA.*
- (v) Volume 11: [Determinations of North American Microlepidoptera, 1881-1884].
- (vi) Volume 12: *List of specimens of microlepidoptera received from William Doherty* (1857-1901) *collected in various localities purchased by Lord Walsingham at an all round price of sixpence a piece.*
- (vii) Volumes 13-14: [British collections purchased by Walsingham].
- (viii) Volume 15: *Collections from various localities given to Lord Walsingham by John Henry Leech (1862-1900).*
- (ix) Volume 16: [List of specimens from H.H. Smith collection

from the West Indies and the Godman and Salvin collection from Central America].

(x) Volume 17: [Notes and determinations of collections including H.J.S. Pryer, E.R. Bankes, T.A. Chapman and O. Staudinger].

(xi) Volume 18: [Walsingham collections from Jamaica, Canaries and Teneriffe].

(xii) Volume 19: [Collecting data from Corsica, France and Spain].

(xiii) Volume 20: [Collecting notes from various countries].

(xiv) Volume 21: Includes a copy of the Walsingham diary for 1871 (original destroyed) of a collecting trip to Oregon, and a register of the Walsingham list of specimen numbers.

(xv) Volume 22: [Collecting data from American specimens].

(xvi) Volume 23: [Collecting data from Algeria and Canaries, 1903-1907].

(xvii) Volume 24: [Notes on the P.C. Zeller collection].

(xviii) Volume 25: [Notes on collections from various countries].

(xix) Volumes 26-27: [Notes on species of microlepidoptera].

(xx) Volume 28: *Index to notebooks labelled Mss by Walsingham, Lord. and Durrant, J.H.*

(xxi) Volume 29: [Catalogue of species of Microlepidoptera].

(xxii) Volumes 30-36: [Walsingham and Durrant determinations].

(xxiii) Volume 37: [Notes on early Microlepidoptera descriptions].

(xxiv) Volume 38: *Notes on specimens in other collections.*

(xxv) Volume 39: [List of species in the H.T. Christoph collections].

(xxvi) Volumes 40-41: [Notes on the O. Staudinger collection and catalogue.]

(xxvii) Volume 42: Annotated copy of Walsingham's *Steps towards a Revision of Chambers' Index with Notes and Descriptions of New Species* (1888).

(xxviii) Volume 43: [Notes on collections sent for loan to British entomologists.]

(xxix) Volume 44: *Notes on the Ms index* [to Hübner's Microlepidoptera in *Sammlung europäischer Schmetterlinge* (1796-1830).]

(xxx) Volume 45: [Notes on Hübner's dates.]

(8) Correspondence collection: Approximately 400 letters, c. 1867-1910, sent to Lord Walsingham and John Hartley Durrant (1863-1928). Includes letters from:

Atmore, Edward A. (1855-1930)	1890-1897
Bankes, Eustace Ralph (1861-1929)	1888-1910
Barrett, Charles Golding (1836-1904)	1868-1899
Buckler, William (1814-1884)	1872-1883
Busck, August (1870-1944)	1900-1908
Butler, Arthur Gardiner (1844-1925)	1881-1916
Constant, Alexandre (1829-1901)	1887-1891
Doherty, William (1857-1901)	1891
Doubleday, Henry (1809-1875)	1867-1868
Fernald, Charles Henry (1838-1921)	1877-1885
Hodgkinson, James B. (1823-1897)	1867-1895
Jordan, Karl Hermann Christian (1875-1972)	1894
Marshall, Guy Anstruther Knox (1871-1959)	1895
Meyrick, Edward (1854-1938)	1889-1894

(9) One Ms volume: *Catalogue of the books presented by Lord Walsingham to the British Museum (Natural History) 1 April 1910 by H. Stringer*, 1914.

334. WALTON, H.J.

Collection

(1) 2 Ms leaves: [Tibet expedition localities and examples of specimen labels]. c. 1905.

Insect Room Lists: Volume 1, item 25.

335. WARREN, Brisbane Charles Somerville (1887-1979)

Born in Fermoy, Ireland, he spent much of his early life in England and Germany. Later he lived in Switzerland until 1934, then returned to live permanently in Folkestone, Kent. From 1902 to 1960 he collected butterflies, building an extensive collection of European Lepidoptera. He published 112 papers and *The Monograph on the genus* Erebia (1936). Warren was a Fellow of the Royal Entomological Society of London and a member of Societas Europaea Lepidopterologica. His collection of 21,000 Rhopalocera and 2,000 slide preparations was bequeathed to The Natural History Museum, London in 1979. B.M.1979-101.

Biographical Reference: 1979. De Worms, C.G.M. *Entomologist's Record and Journal of Variation* **91**:111-112.

Collection

(1) Correspondence collection: Approximately 150 letters between Warren and Entomology staff of the Natural History Museum, London, 1927-1935. This includes correspondence regarding *The Monograph on the genus* Erebia (1936) and the donation of Warren's collection.

336. WARREN, William (1839-1914)

Born in Cambridge, he graduated from the University there and became a school master. He also became a keen insect collector and travelled widely in the British Isles. His main area of study was the Microlepidoptera of Britain. He worked on the Geometridae at The Natural History Museum, London and the Zoological Museum, Tring in Hertfordshire.

Biographical Reference: 1914. Anon. *The Entomologist's Monthly Magazine* **50**:294-295.

Collection

(1) 60 boxes of Mss: Mostly concerned with British Geometridae.

337. WATERHOUSE, Charles Owen (1843-1917)

The eldest son of the natural historian G.R. Waterhouse, Charles joined the British Museum as entomologist in 1866, becoming Assistant Keeper in 1905. He published over 200 papers, including *Illustrations of typical specimens of Coleoptera in the collection of the British Museum, Part 1 Lycidae* (1879) and *Aid to the Identification of Insects*, 2 volumes (1880-1890). He was a Fellow of the Entomological Society of London and served as President.

Biographical Reference: 1917. Distant, W.L. *The Entomologist* **50**:71-72.

Collection

(1) Correspondence collection: Approximately 200 letters mainly to C.O. and George Robert Waterhouse, c. 1897-1916, from entomologists and naturalists of the time, including:

Aurivillius, Per Olof Christopher (1853-1928)	1891
Burmeister, Herman Carl Conrad (1807-1892)	1862
Candèze, Leon (1863-1926)	1889
Crotch, George Robert (1841-1874)	1870
Enock, Frederick (1845-1916)	1915
Fauvel, Charles Adolphe Albert (1840-1921)	1890
Gervais, François Louis Paul (1816-1879)	1842
Horn, George Henry (1840-1897)	1842
Miall, Louis Compton (1843-1921)	1899
Owen, Richard (1804-1892)	1835-1883
Stephens, James Francis (1792-1852)	1839
Walton, John (1784-1862)	1840

(2) Two Ms notebooks: Includes insect records, mainly of Coleoptera, with notes and letters from other entomologists, 1868-1900.

338. WATERHOUSE, George Robert (1810-1888)

Born in Somers Town, England, he trained as an architect, spending his spare time studying natural history. His particular entomological interest was the Coleoptera. In 1835 Waterhouse was appointed Curator of the Museum of the Royal Institute of Liverpool and then became Curator of the Zoological Society of London. In 1843 he was appointed Assistant in the Geological Department, British Museum. He became Keeper of the Mineralogical branch of the Natural History Department. He published papers on a range of natural history subjects and illustrated many of them himself. Works include:

Catalogue of the Mammalia Preserved in the Museum of the Zoological Society (1838), *A Natural History of the Mammalia* (1846-1848), and *A Catalogue of British Coleoptera* (1858-[1861]). G.R. Waterhouse amassed a large collection of Coleoptera; all the type specimens in the collection were presented by him to The Natural History Museum, London in 1875. Later other specimens of interest were presented by C.O. Waterhouse. G.R. Waterhouse was President of the Entomological Society of London and was an Honorary Fellow of the Zoological Society.

Biographical Reference: 1888. Anon. *The Entomologist's Monthly Magazine* **24**:233-234.

Collection

(1) Loose-leaf Ms: [Notes on Amycterinae (Curculionidae)], with annotations by Kenneth Gloyne Blair.
(2) Three Ms leaves: *List of Coleoptera belonging to J.de la Touche.*
(3) One Ms volume: [Register of G.R. Waterhouse's private collection of Coleoptera.]
(4) Two Ms leaves: *Species of* Hister *from the collection of Mr. Waterhouse, British Museum.*
(5) *Pocket Catalogue of British Coleoptera*, published in 1861, heavily annotated by John A. Power (1810-1886).

339. WEBER, Hermann (1899-1956)

Born in Bretten, Baden, Weber studied at the Universität Tübingen and the Zoologisches Institut. He held several research and teaching posts and in 1950 became Professor of Zoology at Tübingen. Weber's major publications include *Biologie der Hemipteren* (1930), *Lehrbuch der Entomologie* (1933) and *Grundriss der Insektenkunde* (1938). He was an Honorary Fellow of the Royal Entomological Society of London.

Biographical Reference: 1956. Alam, S.M. *Indian Journal of Entomology* **18**:311.

Collection

(1) Loose-leaf Ms: Weber's *Biologie der Hemipteren. Eine Naturgeschichte der Schnabelkerfe*, 1930.

340. WEST, William (1836-1920)

Born in Rotherhithe, London, he worked for 56 years in a brass foundry in Greenwich. His first entomological interest was collecting Lepidoptera but he later turned his attention to the Coleoptera, Hemiptera and Orthoptera. In 1919 he presented his collection of over 3,000 British Homoptera to The Natural History Museum, London. B.M. 1919-339. He was one of the founders of the South London Entomological Society and was appointed its first Curator.

Biographical Reference: 1920. Adkin, R. *The Entomologist* **53**:215-216.

Collection

(1) One bound Ms notebook: *British Hemiptera, Homoptera, 4 cases, record of contents*, 1910.

341. WESTMINSTER SCHOOL, Brazilian Expedition

Collection

(1) One Ms notebook: [Relating to an expedition to Brazil by members of Westminster School in 1963]. Collections were made by N. Pinfield and N.S.B. Tanner. The insect specimens were presented to The Natural History Museum, London. B.M.1963-620.

342. WESTWOOD, John Obadiah (1805-1893)

Born in Sheffield, Westwood was the son of a die sinker. The family moved to Lichfield, Staffordshire and in 1921 Westwood went to London to work as a solicitor's pupil. He qualified as a solicitor, but never practised, devoting himself to entomology and archaeology instead. He published many papers and books, and had a considerable reputation as an entomological artist. His publications include *An Introduction to the modern Classification of Insects* (1838-1840), *Arcana Entomologia* (1841-1845), and *The Cabinet of Oriental Entomology* (1847-1848). He was on the staff of the *Gardener's Chronicle* as an entomological referee. Westwood was President of the Entomological Society of London and an Honorary Fellow of most entomological societies of his period. He was an Honarary MA and Honorary Fellow of Magdalen College.

Biographical Reference: 1893. McLachlan, R. *The Entomologist's Monthly Magazine* **29**:49-51.

Collection

(1) One Ms leaf: Copy of Memorandum from John Westwood whilst at Oxford concerning insect specimens from the Entomological Society, The East India Company and the collections of F.W. Hope (1797-1862) and of Westwood himself. These specimens had been borrowed by William Shuckard (1803-1868) and illegally disposed of. Unwittingly Janson, a dealer, purchased these specimens at auction and later sold them to the British Museum. Westwood purchased some of them back.

Insect Room Lists: Volume 1, item 28.

343. WHALLEY, Paul Ernest Sutton (1930-)

Studied at the University of North Wales, and was a postgraduate student with the Nature Conservancy Council (now English Nature). He then became Government Entomologist in Uganda, and this was followed by 30 years in the Entomology Department of The Natural History Museum, London. Whalley was the author of many papers on moth taxonomy. He also published several popular works on butterflies including *The Mitchell Beazley Pocket Guide to Butterflies* (1981) and *Butterfly Watching* (1980). Whalley is a Fellow of the Royal Entomological Society.

Collection

(1) Two Ms notebooks: *Homoptera Auchenorrycha Collection Catalogue*, listing specimens collected in Uganda, 1956-1958.

344. WHEELER, Leonard Richmond (1888-1949)

Born in Highgate, Wheeler joined the Colonial Service in 1914. During World War I he joined the Royal Flying Corps and served in the Balkans until 1919. From 1921 to his retirement in 1938 he held a series of teaching posts in Malaya, where he collected butterflies. He was a Fellow of the Linnean Society.

Biographical Reference: 1949. Wheeler, D.H. *Proceedings of the Linnean Society of London* **161**:253-254.

Collection

(1) Loose-leaf Ms: *List of Collection of Malayan Butterflies (excluding 'duplicates').*
(2) Loose-leaf Ms: *Revised list of Malayan Butterflies (collected April 1934-March 1938) and a few from Hong Kong (collected 13-14 April 1938 or given 1939).*

345. WHITE, Timothy

Collection

(1) One Ms volume: *Harris's Exposition of English Insects with observations and remarks wherein each insect is particularly described* (1766). [This is a manuscript copy of the text from Moses Harris, *Exposition of English Insects* (1766) with original drawings by T. White.]

346. WHITTAKER, Oscar

A keen Hymenopterist, he collected many species of Serphoidea from British Columbia where he lived. His collection of over 2000 Hymenoptera was bequeathed to the Natural History Museum, London. B.M.1947-212.

Collection

(1) Two Ms notebooks: *Author's key to species* [Includes Serphoidea from British Columbia also notes on Proctotrupoidea].

347. WHITTLE, Francis George (1854-1921)

Born in Highbury, Whittle was employed for many years by the Joint Stock Bank, retiring in 1913. He was a field naturalist but was particulary interested in British Lepidoptera, specializing in the Tineina. His extensive collections were bequeathed to The Natural History Museum, London, and included over 46,000 specimens of Lepidoptera. B.M.1922-79.

Biographical Reference: 1921. Sheldon,W.G. *The Entomologist* **54**:302-303.

Collection

(1) 14 Ms volumes: Entomological notebooks on collecting in Britain, c. 1889-1915.

348. WHYMPER, Edward (1840-1911)

Son of a London wood engraver, Whymper went to Switzerland in 1860 to do a series of drawings. While there he developed a lifelong interest in mountain climbing and natural history. In 1879 he went to Ecuador. In 1892 *Travels amongst the Great Andes of the Equator* was published. As a result of the exploration he was awarded the Gold Medal of the Royal Geographic Society.

Biographical Reference: 1948. Hagen, V.M. *South America: Green World of the Naturalist*. xvii + 398pp. London.

Collection

(1) One Ms file: *Catalogue of collections obtained in North Greenland in 1872*. Contains details of Natural History specimens and their localities.
(2) Loose-leaf Ms: Notes for the 1891 Supplementary *Appendix to Travels amongst the Great Andes of the Equator*. Includes 19 drawings of Coleoptera by Henry Walter Bates (1825-1892).
(3) Unfinished botanical oil painting.

349. WICKHAM, Archdale Palmer (1855-1935)

Wickham was educated at Marlborough and New College, Oxford, and was ordained a minister of the Church in 1880. He was an enthusiastic collector of Lepidoptera.

Collection

(1) Five Ms leaves: *Selangore lepidoptera.* [List of Malayan specimens and data]. Concerns the collection B.M.1908-283.

Insect Room Lists: Volume 2, item 87.

350. WILLIAMS, Harold Beek (1889-1969)

A barrister by profession, he became interested in British Lepidoptera and reared many aberrations. Over a period of many years he donated specimens to The Natural History Museum, London. He was a Fellow of the Royal Entomological Society of London.

Biographical Reference: Daily Telegraph 23 April, 1969.

Collection

(1) Loose-leaf Ts: *Notes on the Genus* Gonepteryx *and on the Life History and Variation of* G. rhamni. Includes original drawings of representatives of the genus.
(2) Ts: *Peroneural Forms.* Includes notes from other authors' papers and original pen-and-ink drawings.
(3) Ts: *The British Tiger and Ermine Moths.* Includes original drawings.
(4) Ts: notes and drawings [on the genus *Colias*].
(5) Ts: [The breeding of *Colias dominula*]. Includes photographic plates.
(6) (i) Ts: [Notes on collections including the J. Forsythe Johnstone collection.]
(ii) Notes and drawings on *Catocala fraxini* in Britain.
(iii) Letters of thanks from The Natural History Museum, London regarding donation of specimens, 1939-1951.

(7) Tss:

(i) *A note on* Maniola jurtina.

(ii) *A note on Gynandromorphs and variation in* Bupalus piniaria.

(iii) *A note on the variation of* Epinephele tithonus.

(8) Ts: Euchloe *Ms, Part 1. The genus* Euchloe.

(9) Ts: Euchloe *Ms, Part II. Genera allied to* Euchloe.

(10) Ts: Euchloe *Ms, Part III.*

(11) Ts notes copied from other authors' papers.

(12) Ts notes: [On hybrids.]

(13) Ts: *Dilina tiliae.*

(14) Ts notes: *Boarmia repandata.*

(15) Ts notes: *Pieris.*

(16) Ts notes: *Callimorpha dominula.*

(17) Ts notes: [Temperature experiments].

(18) Ts: *Boarmia rhomboidaria.*

(19) Ts: Notes and Arctiidae plates.

(20) Miscellaneous Ts notes.

351. WILSON, John

Collection

(1) Seven Ms volumes: Includes drawings and photographs of British and foreign Lepidoptera, c. 1940s-1960s.

352. WILTSHIRE, Edward P.

Educated at Cheltenham College and the University of Cambridge, he entered the Consular Service in 1932. However, he has been an enthusiastic entomologist since his schooldays and has published a considerable number of papers on the Lepidoptera. In recent years he has contributed many papers to the *Fauna of Saudi Arabia.*

Biographical Reference: 1991. Wiltshire, E.P. *Entomologist's Record and Journal of Variation* **103**:63-68,119-124,175-180,251-256,307-312; 1992. Witshire, E.P.

Entomologist's Record and Journal of Variation **104**:33-38, 71-74, 149-154, 203-206.

Collection

(1) Ts: *Neotropical insect notes II: Biological observations of insects dependant on Imbauba* (Cecropia) *(Moraceae) in south-east Brazil.*
(2) Ts: *Further notes on the phenological classification of Palaearctic Lepidoptera.*
(3) Ts: *Further proof of protection afforded by Azteca ants to the Imbauba* (Cecropia) *trees of Brazil against insect parasites*, c. 1954.
(4) Miscellaneous Mss, typescripts and photographs.

353. WOLLASTON, Thomas Vernon (1822-1878)

Born in Lincolnshire and educated at Cambridge, Wollaston was forced by ill health to spend winters in Madeira, where he collected extensively. He published over 40 papers but his most important work was *Insecta Maderensia* (1854). His insect collections from Madeira, The Canary Islands, Cape Verde Islands and St. Helena were acquired by the British Museum.

Biographical Reference: 1878. Westwood, J.O. *Proceedings of the Entomological Society of London* **1877**:xxxviii.

Collection

(1) One Ms notebook: *Numbers relating to the localities of the Madeira insects in the British Museum.*
(2) One photocopy Ms notebook: *Reference to the insects in the Madeiran collection formed by T.V. Wollaston.* Concerns a collection formed by Wollaston, presented to the Hope Entomological Collections, Oxford, 1860 and catalogued by J.O. Westwood.

354. WOODHOUSE, Lionel Gilbert Ollyett

Woodhouse was a Fellow of the Royal Entomological Society and Surveyor General of Ceylon during the 1940s. His major work, *The Butterfly Fauna of Ceylon* (1942) was co-authored with G.M.R. Henry. Many of the plates in this work were photographs of butterflies which had their wing scales transferred onto gummed paper by a method known as nature printing. In 1952 he presented to The Natural History Museum, London his collection of Ceylon butterflies containing nearly 4,000 specimens. B.M.1951-301.

Collection

(1) Collection of over 1,000 nature printed butterfly specimens used in the production of *The Butterfly Fauna of Ceylon* (1942).
(2) Correspondence collection, c. 1936-1962: Letters between Woodhouse and the Entomology Department of The Natural History Museum, London mostly regarding the production of *The Butterfly Fauna of Ceylon* and subsequent editions, and the presentation of his specimens.

355. ZELLER, Philip Christoph (1808-1883.)

Born at Steinheim, Würtemberg, Zeller was interested in entomology from an early age. He graduated from the Gymnasium and University of Berlin and studied Coleoptera, Diptera and Lepidoptera during his leisure time. He became a world authority on Microlepidoptera, publishing papers in *Isis*, *Linnaea Entomologica* and *Stettiner Entomologische Zeitung* among others. He travelled to England, where he met many British entomologists, including H.T. Stainton, Henry Doubleday and J.W. Douglas. He assisted in the preparation of Stainton's *Natural History of the Tineina* (1855-73). Zeller's collection of over 31,000 specimens, came to The Natural History Museum, London 1883-1884. His manuscript and correspondence collections came to the Natural History Museum as part of the Stainton bequest.

Biographical Reference: 1931. Essig, E.O. *History of Entomology.* vii + 1029pp. (pp. 808-810). New York.

Collection

(1) 18 boxes of correspondence containing approximately 3,000 letters from international entomologists to Zeller, including letters from:

Bachstein	1877
Barrett, Charles Golding (1836-1904)	1871-1874
Bastelberger, Max Joseph (1851-1916)	1875?
Bellardi, Luigi (1818-1889)	1860
Boheman, Carl Heinrich (1796-1868)	1846-1852
Boll, Jacob (1828-1880)	1873-1877
Brischke, Carl Gustav Alexander (1814-1897)	1877-1879
Brunner von Wattenwyl, Carl (1823-1914)	1876
Burgess, Edward (1848-1891)	1868
Butler, William	1871
Chambers, Victor Tousey (1831-1883)	1876
Christoph, Hugo Theodor (1831-1894)	1860-1882
Dahlbom, Anders Gustav (1806-1859)	1842-1847
Dewitz, Herman (1848-1890)	1880
Dietze, Karl (1851-1935)	1868-1880
Dimmock, George (1852-1930)	1880
Dohrn, Felix Antoin (1840-1909)	1871-1878
Dunning, Joseph William (1833-1897)	1873
Edwards, William Henry (1822-1909)	1871-1873
Eppelsheim, [Eduard] (1837-1896)	1871-1872
Erschoff, Nicolas Grigorevich (1837-1896)	1872-1877
Fernald, Charles Henry (1838-1921)	1878-1879
Fish, Charles	1878-1880
Friedrich, A.	1873
Fritzsche [John]	1864
Fuchs, August (1839-1904)	1872-1887
Geldart, E. M.	1863-1868
Gerstaecker, Carl Eduard Adolph (1828-1895)	1864-1872
Gregson, Charles Stuart (1817-1899)	1869
Grentlenberg, Robert (1823-1886)	1860
Grote, August Radcliffe (1841-1903)	1873-1874
Haag, J.	1872
Haas, Andreas	1872
Haberhauer, J.	1870
Hagen, Herman August (1817-1893)	1871-1877

Harper-Crewe	1862
Hartig, Robert (1839-1901)	1879
Hedemann, Wilhelm von (1836-1903)	1872-1882
Heinemann, Hermann von (1812-1871)	1862-1871
Heller [C].	18[7]8
Heuacker, Wilhelm	1872
Heylaerts, [H.J.]	1874-1883
Hofmann, Fredrich (1798-1870)	1867-1870
Hopffer, Carl Heinrich (1810-1876)	1862-1875
Hornig, Johann von (1819-1886)	1879-1880
Jordan, Robert Coane Roberts (1825-1890)	1869-1873
Judeich, Johann Friedrich (1828-1894)	1875
Katter, J.	1873
Keferstein, Georg Adolf (1793-1884)	1873
Killias, Eduard (1829-1891)	1876-1883
Knaggs, Henry Guard (1832-1908)	1869
Koch, Gabriel (1807-1881)	1879
Kühn, Julius (1825-1910)	1877-1879
Kuwert, August Ferdinand (1828-1894)	1873
Libbach, A.	1864
Lintner, Joseph Albert (1822-1898)	1877
Loew, F. Hermann (1807-1878)	1842-1871
Luben, Alihers	1845
McLachlan, Robert (1837-1904)	1864-1881
Mann, Joseph Johann (1804-1889)	1860-1880
Mead, Teodor Luqueer (1852-1936)	1878
Ménétries, Edouard (1802-1861)	1860
Meske	1876
Meyrick, Edward (1854-1938)	1880-1882
Moschler, H.B.	1870-1882
Muller.	1876
Nicelli, Gustav	1849-1850
Nowicki, Maximilian Sila (1826-1890)	1861
Nylander, William (1822-1899)	1847
Öken, Ludwig Lorenz (1779-1851)	1837-1850

Osten-Sacken, Charles Robert (1828-1906)	1879-1880
Packard, Alpheus Spring (1839-1905)	1863-1875
Pagenstecher, Arnold (1873-1913)	1876-1882
Pergande, Theodore (1840-1916)	1876
Petersen, Wilhelm C. (1854-1933)	1875-1887
Plötz, Carl (1813-1886)	1879-1881
Ragonot, Emile Louis (1843-1895)	1875-1881
Reichlin-Meldegg, Gustav (d. 1880)	1867
Riley, Charles Valentine (1843-1895)	1869-1882
Ritsema, Conrad (1846-1929)	1876
Roessler, [D].	1860-1882
Rogenhofer, Alois Friedrich (1832-1897)	1867-1878
Saalmüller, Max (1832-1890)	1880
Sauter, Anton Eleutherius (1800-1881)	1868
Schilde, Johannes Gustav (1839-1888)	1871-1880
Schlectendal, Doederich H.R. von (d. 1916)	1877
Schmeltz, Johannes Dietrich Eduard (1839-1909)	1871-1874
Schneider, William Gottlieh (1814-1889)	1853-1860
Schøyen, Wilhelm Maribo (1844-1918)	1875-1880
Semper, Johann Georg (1837-1909)	1875
Siebold, Carl Theodor Ernst von (1804-1885)	1851-1870
Sientenis, Franz (1836-1911)	1875-1880
Snellen, Pieter Cornelius Tobias (1832-1911)	1866-1882
Snellen van Vollenhoven, Samuel Constant (1816-1880)	1866
Speyer [Adolph] (1812-1892)	1849-1879
Stainton, Henry Tibbats (1822-1892)	1873-1883
Standfuss, Gustav (1815-1897)	1849-1880
Staudinger, Otto (1830-1900)	1855-1881
Stierlin, Wilhelm Gustav (1821-1907)	1875
Stoll, Otto (1849-1922)	1867-1870
Strange, A.	1857-1866
Strange, G.	1879-1883
Suffrain, Christian Wilhelm Ludwig Eduard (1805-1876)	1868
Taschenberg, Ernst Ludwig (1818-1898)	1862-1870
Teich, Karl August (1838-1908)	1867-1883

Tengström, Johan Martin Jakob von (1821-1890)	1863-1870
Thurau, Friedrich (1843-1913)	1879
Torge, Otto	1879
Trapp, Hermann	1875
Wallengren, Hans Daniel Johan (1823-1894)	1872
Walsingham, Thomas de Grey (1843-1919)	1879-1882
Werneburg, Adolf (d. 1886)	1861-1872
Wocke, Maximillan Ferdinand (1820-1906)	1876-1883
Zaddach, Ernst Gustav (1817-1881)	1876
Zetterstedt, Johann Wilhelm (1785-1874)	1849
Ziegler, Daniel (1804-1876)	1865

(2) 19 entomological notebooks: Include drawings and observations, 1813-1883.

(3) Four Ms notebooks: Contain copies from other authors' entomological works published between 1830 and 1865.

(4) Lists and indexes to the Zeller Microlepidoptera collection in The Natural History Museum, London.

(5) Author's copy of *Chilonidarum et Crambidarum genera et species* (1863). Includes Zeller's manuscript and original drawings by G. Schulz. This item was later published in *Horae Societatis Entomologicae Rossicae* (1881). pp. 154-256.

(6) Zeller's translation into Latin and German of part of H.T. Stainton *The Natural History of the Tineina*, 1855-1873.

Name Index

Note This index lists all significant names which appear in the collections. References are to collection numbers, not pages. Main entries appear in bold.

Geographical Index

Note Significant place names referred to in manuscript collections have been indexed under broad geographical headings. Place names are generally those in use at the time the manuscript was prepared. Most collections have been indexed to the level of country. For Great Britain, counties have also been included. Reference is to the collection number.